T0208619

Scientific Peer Reviewing

Peter Spyns • María-Esther Vidal

Scientific Peer Reviewing

Practical Hints and Best Practices

Peter Spyns
Dept. of Economy, Science
and Innovation
Flemish Government
Brussels
Belgium

María-Esther Vidal
Computer Science Department
Universidad Simón Bolívar
Baruta, Caracas
Miranda
Venezuela

ISBN 978-3-319-25083-0 ISBN 978-3-319-25084-7 (eBook)
DOI 10.1007/978-3-319-25084-7

Library of Congress Control Number: 2015958088

Springer Cham Heidelberg New York Dordrecht London

Printed on acid-free paper

Springer International Publishing AG Switzerland is part of Springer Science+Business Media (www.springer.com)

Preface

Problem Statement

Although peer reviewing forms an integral part of science as a quality control and assessment process, hardly any formal education and training material can be found—unlike for the related activity of scientific writing and presenting. Training mostly happens "on the job" and results after some time in mastering some widely shared good practices and acquiring an implicit "value reference framework" that is more or less common to mostly all researchers.

Even if some literature discusses peer reviewing from a theoretical point of view (pros and cons of the "process" as such, quality issues, etc.), until now very few practical guidelines or introductions are available for young or starting researchers[1] who are looking for some background material on peer reviewing (as already is available, e.g., on how to prepare and write a Ph.D. thesis). Most of the material available covers the medical domain (see Appendix). In particular, as—at least in the European Union—more emphasis is being put on innovative doctoral training, a practical introduction on peer reviewing is useful.

Some information can be found on blogs or websites of editorial companies (but mainly on how the reviewing process for the in-house journals is organized or as instructions/background information to editors[2]). Their target group is not so much potential reviewers but rather submitting authors to whom it is explained how the peer reviewing process works in a concern of transparency and clarity. Even though, as a beneficial side effect, this information and supporting training sessions can help to acquire good reviewing skills, a structured approach to actual peer

[1] For example, Springer provides an online course (http://academy.springer.com/peer-review-academy) but it is password protected.

[2] For example, https://www.springer.com/gp/authors-editors/editors/publishing-ethics-for-journals/4176

reviewing practice for young or starting researchers and Ph.D. students is still largely lacking.

Book Content

This booklet aims to tackle this problem by providing a practical introduction to the practice of peer reviewing. Although it mainly focuses on paper reviewing for scientific events in the domain of computer science and (business) informatics, many of the principles, tips, tricks, and examples are generalizable to journal reviewing and other scientific domains. Some of the principles and tips can also be applied when reviewing proposals for research projects or grants. In addition, many aspects of this booklet will also benefit authors of scientific papers (even outside computer science) as they will gain more insight into how papers are reviewed and hence where they have to pay attention to when writing their papers.

This volume is organized as follows.

- The first chapter contains a short, broad perspective on science policy, discussing why peer reviewing is considered as a quality control instrument of scientific activities. The most prominent advantages and flaws of the peer reviewing practice as well as some recent attempts to organize peer reviewing in a different way are described.
- The second chapter elaborates on the main principles a good reviewer should adhere to, namely, honesty, objectivity, fairness, and confidentiality. Subsequently, a list of specific items concerning a scientific paper that a good reviewer has to examine is provided with accompanying explanations. Also, the most important aspects of personal attitude that a good reviewer should pay attention to when writing his/her review are discussed.
- The third chapter contains a series of (anonymized) real-life examples of actual reviewing practice. The examples illustrate practical tips and tricks regarding the most common "do's" and "don'ts" of peer reviewing. Each example is accompanied by comments and some anecdotal material to explain why the example should be considered as a "do" or rather a "don't." All the examples are review excerpts of actual submissions to conferences or workshops. As journal reviews are much more detailed and specific, they are much less suited as generalizable examples.
- The appendix includes selected references to papers discussing the practical aspects of the peer reviewing process as well as some important articles on more theoretical aspects of peer reviewing. For all papers, a link to the online available publication was added.

Acknowledgments

We would like to express our gratitude to Springer for their support in producing this volume. In particular, we would like to thank Dr. Aliaksandr Birukou who invited and inspired us to write this booklet. We are also grateful to Ralf Gerstner who provided valuable suggestions and guided us through the editorial process.

In addition, we would like to acknowledge our "yearly colleagues for a few days" and other participants at the OnTheMove Academy,[3] which is the Ph.D. symposium or doctoral consortium associated with the OnTheMove Federated Conferences.[4] They all contributed without knowing it to this volume by providing us inspiration through their reviews (the OTMA program committee), comments (the OTMA faculty), and the OTMA review exercise (OTMA participating Ph.D. students). During the OTMA review exercise, the participating students review each other's submissions. These student reviews are commented on during the OTM Academy (OTMA) to train the students in how to peer review. This helps them improve their papers as well. This volume can thus be considered as the extended "printed course text" of the OTMA review exercise session.

Finally, we express our thanks to the general chairs of the OTM conferences, in particular Em. Prof. Dr. Robert Meersman, as they are still willing to support the OTM Academy (even after having provided support for more than 10 years).

We wish you an interesting, inspiring, and above all useful read. Anybody willing to share his/her experiences, examples, tips, etc., is invited to contact us at Peter.Spyns@gmail.com.

Brussels, Belgium Peter Spyns
Baruta, Caracas, Miranda, Venezuela María-Esther Vidal

[3] The OTMA (http://www.onthemove-academy.org/) is a highly tutored and interactive (some might call it "shepherded") workshop where Ph.D. students present their work, receive critical comments and tailored suggestions on their paper and poster and writing and presentation skills, and participate in a peer reviewing exercise. Background material (e.g., general tips and tricks, an example poster template) as well as a comparison with other Ph.D. workshops, are available (cf. http://www.onthemove-academy.org/index.php/otma-downloadables).

[4] www.onthemove-conferences.org

Contents

Chapter 1
The Concept of "Peer Reviewing"

1.1 What Is Peer Reviewing About?

1.1.1 The Role and Function of Peer Reviewing

Scientific activities and scientific endeavors at regular points in time have to prove their "reason of existence": why would anybody put time and money into some aspect related to science and continue to do so, and how is one able to determine the most effective methods, the "best" results, and the most likely promising research topics or directions? The basic idea is that only fellow researchers or peers are able to perform this kind of quality control, as they are the only ones who are equally intensely involved in the subject or sufficiently knowledgeable to redo experiments and falsify/verify the original outcomes or the research hypotheses.

Peer reviewing serves to guard the quality standards of scientific activities at several points in time and acts as a particular form of social pressure and self-regulation through reputation management. At the start, usually a project or grant request is only awarded funding after a positive peer review. Subsequently, at regular intervals (e.g., two-yearly or halfway through the project), an intermediary review determines whether or not the activities are on track. At the end, a final evaluation tries to assess to what extent the goals have been achieved. At each stage, researchers may submit papers for publication, which are again reviewed by their peers for inclusion in a journal or book or for presentation during a conference or workshop, and further publication in the venue proceedings. Each time, peers are the persons who examine and assess the quality and validity of the submissions.

P. Spyns, M.-E. Vidal, *Scientific Peer Reviewing*,
DOI 10.1007/978-3-319-25084-7_1

Peer reviewing is thus not only an important external quality control process, but it is also one of several ways to prevent scientific fraud.

1.1.2 The Importance of Peer Reviewing

As peer reviewing is an integral part of a scientific career, one would expect that Ph.D. students receive training in the matter, just as nowadays courses on scientific writing and presenting are provided in the context of innovative doctoral training. However, just as earlier on a Ph.D. student acquired scientific writing skills through on-the-job training, peer reviewing still seems to be lacking in the majority of the formal doctoral training schemes. Many research groups frequently organize internal paper discussion sessions during which members of the research group discuss new and important articles in their field. These sessions are the closest related activity to peer reviewing, but usually these sessions are not really focused on the process of peer reviewing as such but rather on the advance of the state of the art.

Peer reviewing primarily is a free service to the scientific community, submitters, and reviewers as no direct benefits or compensations are offered to a reviewer.[1] But it does benefit a reviewer albeit indirectly: by examining how others tackle a problem, write a proposal, organize a study, report in a certain manner, or develop a scientific argument, a reviewer is able to learn and improve his/her insights and expertise. Furthermore, peer reviewers have to examine closely within a given time frame a new paper, an innovative proposal, or a challenging grant request and produce a coherent and solidly founded assessment and recommendations for improvements. This activity motivates reviewers to improve their analytical skills and expand their vision of a particular domain of knowledge. As a result, reviewers will write better scientific papers, proposals, and reports.

Peer reviewing is also important from a policy point of view. It functions as a quality assurance process: peers screen and review the intrinsic quality of items submitted for screening so that only excellent ones are accepted. As an instrument for quality control, it not only concerns papers or proposals in the scientific domain but also applies to other themes, such as institutional policies. For example, innovation and science policy and instruments can be

[1] Exceptions are proposal reviews, as, e.g., organized by the European Commission that usually foresee in some form of financial compensation for the time and travel (if applicable) spent.

reviewed by peers. Peers are then considered in their broadest sense, namely, persons active on the same topic (e.g., civil servants of one country assess the science policy of another country with an advisory purpose). Peer reviewing has outgrown its strictly scientific context and has become a general assessment principle.

One of the priorities of the European Research Area, namely, efficiency of research systems,[2] relies on peer reviewing of research proposals as an instrument to organize competitive funding with the aim to raise the level of excellence. Hence, young researchers, who later on in their career might be invited for evaluation panels and program committees, should become familiar with peer reviewing and receive some form of structured training in the matter.

1.2 How Does Peer Reviewing Work?

The main purpose of peer reviewing is to deliver an objective and neutral assessment of a submission that is shared by more than one reviewer and that covers multiple points of views. The traditional way of performing peer review is called "blind" (cf. Sect. 1.2.1) and presents advantages (cf. Sect. 1.2.2) as well as disadvantages (cf. Sect. 1.2.3). Some new trends aim to maintain the strong points of peer reviewing while reducing its drawbacks (cf. Sect. 1.2.4). Modern social media technologies facilitate these new trends.

1.2.1 Blind Reviewing

Typical of the "regular" reviewing process is that the identity of the reviewers is not disclosed to the submitters. Reviewers can anonymously express their opinion in "all freedom" without having to bother with potential "fallout" or some form of retaliation.[3] To prevent reviewers from

---

[2] The Council of the EU has endorsed the European Research Area Roadmap (http://data.consilium.europa.eu/doc/document/ST-1208-2015-INIT/en/pdf) adopted by the European Research Area and Innovation Committee that explicitly promotes the application of international peer reviewing as an important instrument for an effective national research system (priority 1): ERAC 1208/15, p. 6, 20/04/2015.

[3] Reviewers might fear that their submissions might be negatively evaluated by authors of submissions rejected by the reviewers if the authors know their identity.

abusing their position and to guarantee a variety of points of view as well as a sufficiently consensual final opinion, usually three or an odd number of reviewers examine the same submission. If the submission includes the identity and affiliation of the submitters, the reviewing process is called "single blind." If the identities of both reviewers and submitters are not disclosed, the reviewing process is called "double blind." Double-blind reviewing is considered as an extra guarantee that reviewers operate with an open mind and are unbiased toward the identity of the submitters. For example, reviewers sometimes hesitate to assess negatively or reject submissions by well-reputed researchers or researchers they (personally) know well.

Another measure to enhance the objectivity and neutrality of peer reviewing is to hide the reviews and/or identity of the reviewers to the other reviewers. Again, the underlying principle is that a reviewer performs the review in an independent and autonomous manner without being influenced by other reviews beforehand (cf. Sect. 3.1.13). Often, once he/she has entered his/her review, a reviewer can have access to other reviews of the same submission. To avoid that reviewers might be "impressed" by reviews of well-known researchers and subsequently adapt their review, the reviewers' identity is not always shared among the reviewers. In case of strongly diverging or opposed opinions, a chair or editor can organize a (moderated) discussion among those reviewers. It is the choice of the chair or editor to disclose or not their identities to these reviewers.

1.2.2 Arguments in Favor

The most important argument to organize the scientific quality control mechanism as a reviewing process by peers is that members of the scientific community (the peers) are the best to judge the work of a colleague. Usually only other researchers on the same topic or in the same domain have acquired enough expertise and knowledge of the area to be able to judge and assess the work of a colleague. In principle, peers participate in the reviewing process as a service to science in general: no personal benefits are acquired by reviewing.[4]

[4] An exception are project proposal review sessions or project progress evaluations for which reviewers are compensated for—e.g., by the European Commission—as it takes up quite some time from the reviewers.

This basic insight is not limited to some local assessment habit but has been acknowledged worldwide. Peer reviewing is universally well accepted and established as a best practice to screen submissions on their internal qualities. It also involves peers or experts from all over the globe.

This universal dimension adds to the objectivity and neutrality of the reviewing process as it reduces the probability that reviewers have close ties with the submitters. In any case, reviewing happens anonymously (blind or double blind) to facilitate that reviewers perform their review independently. Involving reviewers from all over the world makes it easier to select a sufficient number of well-renowned researchers as reviewers, which increases the good reputation of an event or journal.

It is important to have a sufficiently high number of reviewers available. It is commonly accepted that a review is to be done by at least three reviewers in order to guarantee the objectivity of the reviewing process. Involving a sufficient number of reviewers averages out conflicting opinions or scores and enhances the soundness of the overall review outcome.

1.2.3 Arguments Against

Some of the arguments in favor can be turned around and become important flaws. The most critical argument against the current form of peer reviewing is that it mostly rewards "average or mainstream thinking" as it seeks consensus among reviewers. Disruptive ideas or revolutionary breakthroughs are said often not to be recognized by reviewers. Some conferences and journals have become too connected with a certain scientific community with its internal research dynamics such that other or new "deviant" opinions are not so easily appreciated.

Editorial boards or program committees might become too much of an "in-crowd" so that reviewers might show a too friendly or lenient behavior toward other "incumbents" and might be less receptive toward "outsiders" or newcomers. In particular, when a chair or editor invites his/her "scientific friends,"[5] a positive bias toward an inner-circle or potential conflicts of interest (cf. Sect. 2.1.1.2) are to be avoided. In some cases, in particular in small research communities, it is challenging to find a good combination of "usual suspects" and promising young researchers who cover all topics a

[5] In all fairness, chairs or editors inviting reviewers they are acquainted with is not a problem "per se" as the chair or editor knows the expertise and reliability of these reviewers better compared to other unknown reviewers.

conference or journal (special) issue wants to address. So it can happen that the submitting author(s) is the best expert(s) on the topic at hand. The smaller the community, the faster potential conflicts of interest may arise.

The best known experts are frequently solicited to act as reviewers as they are supposed to assure a high level of quality of the reviewing process leading to a high quality of the accepted submissions. These experts are usually very busy (with reviewing and other academic duties) so they have become quite selective in accepting to become a reviewer. What often happens is that they assign the reviewing task to their Ph.D. students. However, this sometimes goes against the idea of a peer being the most appropriate one to judge the work of another peer as not all Ph.D. students are experts yet (and in many cases even have to learn the "peer reviewing trade"—cf. Sect. 3.1.5). To avoid this situation, some venues explicitly prohibit that reviewers select sub-reviewers to evaluate the papers that have been assigned to them.

Guaranteeing anonymity can lead to unwanted effects. Reviewers can voice inappropriate comments (cf. Sect. 3.1.3) or base their opinion wrongfully on other factors than the mere content of the submission as they feel protected by the anonymity. And, in the case of double-blind reviewing, a good expert can often recognize the submitting author(s).

1.2.4 New Trends

To reduce the existing flaws of peer reviewing and strengthen the overall principle of peer reviewing as a quality control process of scientific activities, several adaptations (and combinations of them) to the peer reviewing process have been proposed and put into practice, some with more popularity and success than others. The availability of new ICT tools and social media technology greatly facilitates modifications in the peer reviewing process.

The most important change is the shift toward what is called "**open reviewing**" or "open evaluation." In essence, open reviewing discards the principle of anonymity. Reviewers can sign their reviews which are public and published on a website. Contrary to the idea that anonymity is beneficial for reviewers freely expressing their opinion, openness is considered to enhance the overall review quality as reviewers publicly, on a voluntary basis, take responsibility for their reviews, which can be "reviewed" in turn by other experts.[6] An open review is the result of an (asynchronous)

[6] The Semantic Web Journal (http://www.iospress.nl/journal/semantic-web/) is a good exemplar of these novel venues.

interactive and entirely transparent process. A particular form of open (and synchronous) reviewing is a "writers' workshop"[7] (combined or not with shepherding sessions—cf. infra) during which authors listen to the comments and suggestions on their paper by workshop participants and integrate them in a subsequent version of their paper.[8]

It has become technically possible to not only perform reviews before the submission is published but also after its publication. A more in-depth and transparent scientific discussion on the content of a paper is started with discussants posting their reactions in all openness on a website. Already now editing houses publish journal papers online, although usually taking the paper format as starting point, with the possibility to add comments online.[9]

Online reviewing and publishing makes it much easier to include extra material with the uploaded paper, e.g., underlying data sets, intermediate results, or stand-alone software components. Available data are not only useful for the readers of the papers afterward but also for the reviewers before the publication has been accepted and made public (open research data). In principle, scientific fraud (tampering with or fabricating raw data, screenshot manipulation, or fallacious demonstrator software) or unintended mistakes in good faith should become easier and faster detectable. Even if reviewers have not enough time to examine all the extra information, readers can step in and perform extra examinations and react accordingly (the falsification principle).

Reviewing papers is no longer a purely scientific matter as peer-reviewed journals that apply the principle of "gold access" (the author pays for the costs of immediate publication[10]) have an economic interest in publishing

[7] For example, http://www.cs.wustl.edu/~schmidt/writersworkshop.html

[8] The LNCS Transactions on Pattern Languages of Programming only accepts submissions that have been reviewed during a writers' workshop.

[9] A step further could be to publish papers as blog entries with moderated discussion threads associated with each paper. Only registered subscribers have access to the information and are allowed to post reactions (reader commentary). Citation impact could be measured by the number of clicks on the paper (positive), reactions and references to it, etc.

[10] The alternative is "the green road to open access" (an author makes the publication available himself/herself) not requiring any payment by the author. More and more it is assumed that some public repository is set up (by a research field, by universities, by the government, etc.) and paid for. This "green access publication" (if the paper was originally published through the regular, commercial "closed" channel) has to respect a publication embargo for a certain period imposed by the commercial publisher.

enough papers to maintain at least a cost breakeven situation. Economic reasoning might take the overhand on scientific reasoning, resulting in more less qualitative papers being published.[11] In that sense, more involvement of readers in checking the quality of the publications is positive.

Other improvements are already now being integrated in the current peer reviewing process. In particular, many conferences and workshops experience a need to stimulate their reviewers in doing a good job. Depending on the size of the conference (and number of submissions), mechanisms are sophisticated and automated or rather impressionistic and manual. Some (large) conferences organize a **meta-review** stage. Dedicated experts only check the quality of the reviews and they can flag a review if they judge that that review does not meet the quality standards (or is lacking). It is then up to the chairs or organizers to decide on what happens with these flagged reviews. Program committee chairs usually do this themselves (and implicitly) if the number of submissions is manageable. They can ask reviewers to rework their comments, override their recommendations, or try to find additional reviewers. An alternative is to divide the domain covered by the call for papers in several subtopics, each having a specific chair and sometimes even a specific program committee. This makes the reviewing process more manageable.

Some conference organizers give authors the opportunity to produce clarifications to questions raised by the reviewers or to correct perceived misunderstandings in the review. During the evaluation, reviewers are therefore invited to include concrete and specific questions and comments. Subsequently, authors are offered the opportunity to reply and clarify the issues and correct factual errors in the review. The reply is used as complementary information to accept or reject a submission. Meta-reviewers are responsible that reviewers take this additional information into account and correlate their scores to the quality of the submitted paper and the newly provided information.

A complementary technique is to allow authors of rejected submissions to start a **redress procedure** (rebuttal stage). Authors can argue about the validity of the reviewer's negative comments and provide arguments to "prove" that the reviewer made a mistake and unjustifiably rejected their submission.

Additionally, borderline papers can go through a **shepherding** process: one of the more experienced reviewers or meta-reviewers works with the

[11] Have a look at http://scholarlyoa.com/publishers/ to know which open-access journals to avoid.

authors on the weak aspects of the paper to improve and resubmit it. The shepherding process can take several iterations, before the shepherd finally decides whether the paper can be accepted (or not).

A new idea is about reusing the reviewing efforts spent on a journal paper by transferring this "older" review to the group of new reviewers when a paper is resubmitted elsewhere (with or without modifications).[12] *eLife*, the *British Medical Journal*, the Public Library of Science, and the European Molecular Biology Organisation have announced a peer review consortium that will try to implement this idea of "portable reviews."[13] An advantage would be that authors receive the comments and the editorial decision much quicker and that the second group of reviewers avoids repeating similar comments as the first ones.

Another way to shorten the sometimes very long reviewing time, in particular for journals, is what is called "**re-review opt out**."[14] Revised papers are usually resent to the reviewers for another round of comments, which may result in yet another set of recommendations for modifications and suggestions for improvements. This can result in a very long-lasting reviewing process, with the risk that the information in the paper already has become obsolete at the date of publication, or that certain suggested modifications cannot be implemented due to changed circumstances. In order to speed up this process, authors of papers that are considered to be "publishable after modifications" can opt out of the next round of reviewing. The reasoning behind the "re-review opt-out" strategy is that authors for their reputation's sake will avoid publishing less qualitative papers and will do their utmost best to deliver a highly qualitative, revised paper. This would avoid another round of reviewing and shorten the time to publication. Nevertheless, the editorial team of the journal still examines the revisions and can reject the revised paper anyhow if needed. Several other new initiatives to renew the process of peer reviewing are under consideration, e.g., in vs. out channel postpublication review or (commercially organized) prepublication review.[15]

---

[12] For conferences, let alone workshops, this seems much more difficult (and probably much less worthwhile).

[13] Personal communication by Ralf Gerstner (Springer)

[14] http://www.biomedcentral.com/content/pdf/1741-7007-11-18.pdf

[15] http://journal.frontiersin.org/article/10.3389/fnins.2015.00169/full

1.3 To Conclude

This chapter described the context of the peer reviewing process, why it is important for scientific quality control and assessment, what are its main advantages and flaws, and how recent trends try to remedy the flaws. With this overview a reader is better prepared for the following, more practical chapters as he/she is able to understand why certain practices are put in the focus and why they are recommended or should be avoided.

Chapter 2
Characteristics of a Good Reviewer

2.1 Some Good Review Principles

Although the principles mentioned below are discussed in the context of the reviewer–editor or reviewer–chair relationship, some of them can also be considered from a reviewer–co-reviewer perspective as many reviewing systems allow reviewers to see the comments of their co-reviewers. In the case of open reviewing (see Sect. 1.2.4), authors and readers also can access all review details including the name of the reviewer. Hence, being an honest and fair reviewer is also a matter of personal reputation management, informal good scientific "behavior," and quality control by peer pressure.

2.1.1 Be Honest About

2.1.1.1 Your Level of Expertise

Most research areas are quite broad and can span many different methodologies, techniques, approaches, or theories. For a conference organizer, it is thus not always crystal clear which members of a program or review committee are the most appropriate ones, even if the committee members have indicated by means of keywords their preferred domains. Conversely, if reviewers are allowed to express their preference to review specific papers, most of the time they have to rely on the keywords provided by the authors or the abstract submitted. In either case, it can happen that a paper does not match the (level of) expertise of its reviewer.

P. Spyns, M.-E. Vidal, *Scientific Peer Reviewing*,
DOI 10.1007/978-3-319-25084-7_2

Therefore, many conference review systems include some mechanism for a reviewer to indicate how well acquainted he/she is with the topic(s) of the paper to be reviewed. Popular labels are "confidence" or "familiarity": the reviewer grades himself/herself on the level of confidence he/she attributes to his/her review as being an appropriate one. Implicitly, a reviewer performs a self-assessment on how knowledgeable he/she is regarding the expertise required to review the paper.

As a good reviewer you should be honest about your expertise and level of expertise. Do not falsely give the impression that you are an expert in the matter if that is not the case. You would be surprised how easy it is to detect so-called or would-be experts when they are not. Remember that other reviewers, meta-reviewers (cf. Sect. 1.2.4), or even the committee chair(s) might be "real" experts and that program chairs or journal editors always compare the comments and scores of the various reviewers. In the case of contradicting reviews for the same paper, meta-reviewers or chairs often organize a discussion round between the reviewers to have their opinions converge or to come to a compromise. It is a mere matter of respect for all parties involved as well as a sign of good and honest academic conduct not to boast your level of expertise.[1]

Some reviewers are afraid to admit that they are less (or not all) suited to review a paper. In particular, when the number of program committees a researcher is invited to functions as a scholarly performance indicator or token of academic prestige, it is tempting to "accumulate" as many program committee memberships as possible without any further consideration. Serious chairs, organizers, or editors usually apply some form of quality control[2] as they strive to send back serious and substantiated reviews to the submitting authors (as an indicator of the quality and reputation of a conference or journal).

The worst that could happen when you indicate that you are only moderately familiar with the paper topics is that the program chair attributes a lesser weight to some of your remarks. Remember that you might be a good expert for the other parts of a paper and therefore still deliver a valuable and fair review. But you will gain his/her esteem and be remembered as an

[1] A notorious example of bad reviewing practice are the paper submissions that have been automatically generated by combining lots of buzzwords without any real research done and that, shamefully, have not been rejected by lazy or incompetent reviewers (cf. Sect. 4.4).

[2] Some large conferences include a meta-review in the overall reviewing process (cf. Sect. 1.2.4).

honest, serious reviewer. And probably he/she invites you again as a reviewer at a next occasion.

2.1.1.2 Your Potential Conflict of Interest

Depending on how strictly or formally reviews for a conference, journal, or project proposal are organized, reviewers are asked to officially declare that no conflict of interest arises. It is not always clear when a reviewer has to admit that a conflict of interest may arise. Typically, it is considered that members of the same research group (or even the same faculty of department) cannot review one another's work. The same goes for researchers who regularly and intensely cooperate (e.g., in the same project). Good program chairs or journal editors try to avoid as much as possible potential conflicts of interest when assigning a paper to reviewers. Nevertheless, they do not always know (or take the trouble to investigate) the connections between researchers. As reviewers can become biased[3] (even if unintentionally) in their opinion on papers, proposals, or reports submitted by persons they know well or cooperate with, it is customary that reviewers formally declare before starting the actual review that no conflict of interest exists. Hence, reviews are void of any interest (except furthering science in general).

If a double-blind review is organized, a reviewer might not always detect from the start that the submitting author(s) is a "close" colleague. If you as a reviewer discover while you are reading the paper or proposal that some form of conflict of interest might arise, you should immediately notify the chair or editor and explain why you think there is a risk of a conflict of interest. Some guidelines do exist. According to the Committee of Publication Ethics (COPE), a conflict of interest arises between the reviewer and reviewee when:

> . . . they work at the same institution as any of the authors (or will be joining that institution or are applying for a job there); they are or have been recent (e.g. within the past 3 years) mentors, mentees, close collaborators or joint grant holders; they have a close personal relationship with any of the authors.[4]

For example, if you, as a researcher, are in competition (a novel research outcome, a new or alternative method, a first publication, or better results)

---

[3] Bias can also result in more negative reviews, if, e.g., close colleagues are in some sort of (internal) competition or conflict situation. A positive bias might lead to a less critical review or too "lenient" scores or comments.

[4] Committee of Publication Ethics (COPE), Ethical Guidelines for Peer Reviewers: http://publicationethics.org/files/u7140/Peer%20review%20guidelines.pdf

with the submitting authors, it is ethical to discuss this openly with the chairs or editors.[5] If he/she judges that indeed a conflict of interest has occurred, you will be asked to refrain from further reviewing the paper while maintaining the usual reviewer confidentiality (cf. Sect. 2.1.3). Again, acquiring credibility takes a long time, while you can lose it very quickly by behaving in a dishonest way. Luckily researchers usually are honest people.

2.1.1.3 Your Personal Involvement

Well-reputed scientists are often solicited for review committees but usually do not have that much time available. It is becoming more and more widespread that very busy reviewers delegate the reviewing tasks to their Ph.D. students or junior collaborators as part of learning the academic job. If you have the luxury of assigning the review task to junior/other colleagues, you should name them as sub-reviewers of the paper. Not only are you rewarding and acknowledging the hard work of your colleague by giving him/her visibility (usually additional reviewers are listed separately in proceedings), you are also not deceiving the chair or organizer who has invited you for your expertise. And finally, it is in your advantage that you clearly identify the author of the review. You definitely do not want to be held responsible for a bad review that is not yours (cf. Sect. 3.1.5). That is also why it is a good practice to always check the review submitted by others under your responsibility (in your name). And for them, it might be a real learning experience if they receive your comments on their reviews.

2.1.1.4 Your Commitment of Timely Delivery

Timely delivery is not only a matter of self-discipline, time management, and work planning but also involves an honest and respectful attitude toward the one who invites you as a reviewer. Do not promise what you cannot achieve. If reviewing six papers within the given time span is probably too much given your overall workload and timing constraints, but three will work fine, simply notify the chair, organizer, or editor. In most cases, he/she will be happy that you can at least review some papers and he/she will be assured that these reviews will be submitted on time.

Reviewers, in particular if they have never acted as chairs, organizers, or editors, are unaware or largely underestimate the hard deadlines imposed on

[5] In any case, do not use the "weapon" of peer review to block your competitor.

these latter by editing or publishing organizations. Usually, one late review only has a very minor impact on the overall event or publication; ten late reviews, on the other hand, do have a substantial impact. In particular, if the reviewers who are late do not respond to reminders by the chairs, the latter become stressed by pondering whether or not they have to look for alternative reviewers.

Accidents or delays by "force majeure" or "force of nature" can always happen. That is a fact of life. However, it is your responsibility as an honest reviewer to notify as soon as possible the chair(s), organizers, or editors if an unforeseen problem has occurred. They will be unhappy as they are faced with a new problem to solve, but they will be less unhappy as they know it on time. Waiting until the very last moment—even with the underlying intention of trying hard to deliver anyhow—to notify the chair of your default only creates frustration at both ends. And avoid using (the same) cheap excuses too often (e.g., some mothers-in-law apparently have a very frail health condition).

A phenomenon that an honest reviewer should absolutely restrain from is accepting to be on the review committee and from then on stopping all communication[6] and not delivering any review at all. One can only guess why some reviewers behave like this (e.g., illness, change of job, too much work, afraid to decline the initial request, pressure of performance indicators, or presumed prestige, deliberately stalling a publication that directly is in competition with their own work). Some of the potential causes are more understandable than others, but nevertheless, chairs usually do not forget this easily.[7]

2.1.2 *Be Objective and Fair*

The purpose of a review is to give an objective and fair appraisal of the paper, proposal, or application submitted for review. Ideally, submitting authors should never be disappointed by the reviewers' comments. If the paper, proposal, or application is accepted, submitters are obviously happy. In the negative case, submitters should at least understand why their submission has been rejected, accept the reasons given, and make use of the comments to improve their submission. In that sense, they should be happy

[6] Some online reviewing systems log the history activity of reviewers so that chairs rapidly see who has never logged in to the system.

[7] Blacklists do exist and are passed around (by word of mouth). Chairs do remove names from the reviewers' lists on the website or proceedings.

to have received objective, external, free, and competent advice, which can help them to improve their work.

It implies that a reviewer adopts a professional and helpful attitude with the ultimate aim to improve the scientific state of the art. Hence, personal emotions or feelings (toward the submitting authors) are to be discarded in favor of substantiated scientific arguments. Not a single submitting author likes to receive a rejection notification containing succinct statements such as "I do not like the approach" (cf. Sect. 3.1.7). Always formulate your comments in a respectful way and maintain a high level of courtesy (no disrespectful nor inflammatory statements—cf. Sect. 3.1.3).

References to philosophical, ethical, political, or religious background of the submitting authors are definitely to be avoided—except if it actually concerns the scientific content of the submission.

Even if a reviewer would never try to solve a problem in the way described in a paper under review, as long as the authors adhere to a solid scientific methodology for tackling the problem and discussing the results, a reviewer cannot simply disqualify the work described in the paper because it is not in his/her personal line of work or thinking. Or it is not because the authors do not know or cite your work as a researcher that the entire paper is bad (cf. Sect. 3.1.9). At least reviewers should be aware of such potential personal bias and control and avoid it as much as possible by giving fair scores substantiated by sufficiently elaborated comments (including references to other work) or motivations (cf. Sect. 3.1.12).

Not focusing too much on less relevant details in a paper or proposal is one appropriate way to provide a fair review. Some reviewers, usually due to a lack of in-depth expertise, tend to focus on the parts they do master. In the extreme case, this leads to review comments on language style, paper structure, or isolated issues without many comments on the overall scientific content. Serious editors, program chairs, or workshop organizers easily spot this behavior. Moreover, as more and more submitting authors are asked to provide an accompanying letter detailing their reaction on the reviewers' comments, they make use of these possibilities to introduce a rebuttal request or simply voice by e-mail to editors or chairs their discontent with unprofessional and unfair reviewers. So if you, as a reviewer, discover that certain, important parts of a paper are outside your expertise, you have to notify the chairs or editors (see Sect. 2.1.1.1).[8]

[8] Almost all reviewing systems or forms include a section "comments to the chair or editor." If a paper contains parts for which you are not the most appropriate reviewer, you can mention this in that section.

2.1.3 *Respect Confidentiality*

In the context of a (double) blind review, reviewers are not supposed to disclose their identity toward the submitting authors. Some reviewers would like to contact authors to obtain more clarifications or explanations about parts of the submitted paper, proposal, or grant application. However, if decided by the editors, chairs, or organizers that reviewers cannot unilaterally and directly contact submitters, reviewers should respect this decision. If not, submissions are not treated on the same foot—only some privileged authors would be given the possibility to influence certain reviewers who might adapt their judgment based on information not contained in the paper, proposal, or application (or even not related to it, in the extreme case leading to scientific fraud).

Confidentiality also applies to the content you have been given access to as a reviewer. Normally, you would only be able to learn about the content of the paper, proposal, or report at the time of publication. As a reviewer you have access to knowledge not yet generally available. Nevertheless, you cannot use, let alone disseminate, this information for other purposes than the review at hand (even after the end of the reviewing process). You should thus avoid being intellectually "stimulated" or influenced by certain ideas in a submitted paper to use them for your proper work (as it can happen that the paper may end up not being published at all).

In particular, when reviewing project proposals reviewers usually have to sign a nondisclosure agreement. Confidentiality is usually implicitly assumed for the paper reviewing process. Keeping confidentiality, even if not explicitly asked for, is an important aspect of overall research integrity.

2.2 Issues to Pay Attention To

The following sections provide various, specific items to consider when writing a review. Some are linked to the paper reviewed (the object). They constitute a kind of checklist to follow during the reviewing process. Some of the items can even be used as paper writing guidelines.

Others are related to how a good reviewer (the subject) performs his/her review. The review principles presented in the previous sections are generally applicable, i.e., also to project proposals, grant applications, journals, conference and workshop papers, and even other kinds of reviews. In the following sections, however, the focus lies on scientific paper reviewing.

2.2.1 The Review Object (Paper Under Review)

A scientific paper is a good scientific paper for several reasons. In general, a reviewer gives an overview of the strong and weak points of a paper. First and foremost, the content must be innovative and improve the state of the art, but also both the applied methodology and evaluation method need to be sound and rigorous. In addition, the paper must be well structured and linguistically correct to adequately convey its content to readers. Consequently, a good reviewer should pay attention to all these aspects (and more) to produce a thorough review.

Usually conference organizers, program chairs, and journal or book editors provide a list with the most important review criteria—e.g., novelty, potential impact, scientific excellence, or technical depth in line with the major aims of the conference, journal, or book. Depending on the layout of the review form, specific comments are to be given for each criterion as an explanation of the score(s) given by a reviewer. Alternatively, one general freestyle comment field is available to a reviewer to enter any type of comment. Whatever the case, good and experienced reviewers also systematically apply their own set of additional criteria in order to come up with a review as complete as possible. You can use the items discussed below as a checklist or structure for your own reviews (cf. Sect. 3.2.3).

- Title and Abstract

  A title should adequately "flag" the content of the paper. Although a title may be catchy, a scientific paper is not a newspaper article. A potentially interested scholar wants to decide from the title whether or not he/she spends time on reading the paper. Also, most of the search engines display paper titles in their result list. Hence, the precise wording used in the title is very important and deserves proper attention and thinking. A reviewer is one of the first external readers to judge whether or not a title is appropriate. It happens that authors make their title more appealing, promising more than what actually is conveyed in the paper. A good reviewer should point this out. The best moment to assess the adequateness of a title is after having read the entire paper.

  An abstract functions roughly similarly to a title, but with more lines of text. Some journals impose a fixed structure for the abstract (purpose, problem, method, material, and results) and usually limit it to a maximum number of words. A good abstract is a stand-alone summary of the paper. All the core elements of a paper should be included somehow. Abstracts are a not a mere preliminary introduction or a repackaged conclusion section. As an abstract is a section on its own (and sometimes even published on its own), it must be a self-containing text (no abbreviations, no references, no URLs, no undefined concepts, etc.) with its internal logic or flow of thought. Less adequate abstracts are a concatenation of copied/pasted phrases from all over the paper. Again, an abstract is best judged after a reviewer has read the entire paper.

- Motivating Example or Use Case

  It is important that authors identify scenarios, real or synthetic, describing or exemplifying the problem at hand that at that moment existing approaches fail to solve or tackle appropriately. Motivating examples can be included in the introduction of the paper or constitute a separate section. A good reviewer has to assess whether the motivating example illustrates a real use case, check the drawbacks of the current approaches to the problem, and examine the feasibility of the new approach to provide a better solution.
- Research Questions and Potential Impact

  Another frequently used general criterion to be evaluated is the importance of the research questions or, alternatively, the potential impact of the work. As this criterion is more of a speculative and subjective nature, it can sometimes be difficult to provide an objective and substantiated answer. A well-written paper with a sound methodology can be "killed" if considered as marginally relevant to the field or having only a superficial potential impact. In essence, a negative comment on this criterion to authors could mean that they have spent a considerable amount of their research time on futilities. Reviewers should thus formulate a negative opinion on this criterion in a cautious way backed up by references underpinning this opinion.
- Proposed Approach and Properties

  A good reviewer should assess if the problem to be solved is clearly defined using an existing formalism and whether its theoretical properties are stated. For example, in computer science time and space complexity are important aspects of a problem. Also the solution(s) or improvements as well as their properties should be expressed in terms of a logical or mathematical framework. The proposed approach should be positioned with respect to existing state-of-the-art approaches.
- Evaluation Methodology

  Many papers only superficially describe the evaluation method(s) used. A common flaw is that authors who usually design and implement a software system are the same ones who design the evaluation and test procedures and are the same (and only) ones who subsequently execute them, regularly even without explicitly mentioning this. Although in practice it is sometimes impossible to avoid this kind of potential bias, at least this situation should be mentioned as it may compromise the testing and evaluation results. Reviewers have to be aware of these and other types of potential biases and, if needed, make authors aware of them—in particular if authors present their results in an overly positive way. A detailed description of the hypotheses of the evaluation, benchmarks, baseline approaches, gold standards, evaluation methods, procedures, and analysis of the observed results is required. Statistical methods should be used to ensure reproducibility, significance, and generality of the reported experimental results. Further, a good reviewer must be prepared to not only evaluate positive results but also results that falsify the hypotheses, i.e., negative results.

- Future Directions and Valid Conclusions

  Rarely a matter under research can be convincingly "closed" by a single publication. Usually, results lead to new avenues for future research. Papers without some statements (sometimes in the conclusions section) on future work lack ambition and point to an insufficiently elaborated paper. Reviewers have to challenge authors for their new ideas resulting from their current work. Kind reviewers might even give some suggestions for future work themselves, although authors cannot expect this to be a standard attitude by reviewers as the latter cannot be credited for it (or only anonymously).

  A conclusion summarizes what authors consider to be their most important messages of the entire paper, not necessarily only of the results obtained. Some authors—willingly or not—overgeneralize the importance of their work, minimize the limiting factors, or forget to mention risks on biases—e.g., they do not repeat that their experiments, study, and evaluation happened in a very controlled environment. Reviewers need to be the guardian angels of "time-pressed" readers who only read the introduction and the conclusion sections. A reviewer should carefully examine whether the conclusions put forward are really substantiated by the paper.

  Conclusions echo the statements made in the introduction as the conclusion section is supposed to summarize the "answers" to the research questions and challenges mentioned in the introductory section.

- Quality of Bibliography

  A good list of bibliographic references complies with several criteria. It contains references to the most relevant related work and state of the art at the time of writing. The references are complete (full name of the source and page numbers), correct (no typos in the names), and appropriately formatted (according to the style guide)—cf. Sect. 3.1.9. As a reviewer, you should check whether the bibliography represents a good mix of older and more recent work. If the most recent references are already some years old, it is an indication that the paper has been lying somewhere on a shelf or has been submitted and rejected earlier. References to very general textbooks rather than to thematically focused publications can point not only to a very new topic or area but also to authors who are new to a domain or unaware of previous work done. The same goes for a list of only or majorly recent references.

Although self-referencing cannot be excluded (as every researcher builds on his/her previous work), authors should not exaggerate, as it can also be a sign of a researcher working in isolation without considering related work in his/her area. It should be clear that reviewers themselves have to be seasoned researchers in their domain and well aware of the latest work by peers and new trends in their field.

- Language and Style

  It is delicate for a reviewer, in particular as a nonnative speaker, to make remarks on language and style errors. Nevertheless, fuzzy language obfuscates the messages an author wants to convey. Authors should make use of available spell and style checking tools. A reviewer should not do all the language/stylistic corrections for the authors. But reviewers cannot simply overlook (evident) linguistic and stylistic problems and have to point out sentences that are difficult or ambiguous to understand. This evidently also applies to review comments (cf. Sect. 2.2.2).

- Paper Structure

  Some conferences and journals (also depending on the discipline) impose a fixed paper structure with predefined headers (e.g., context, related work, material and methods, results, evaluation, discussion, future work, conclusions). This may seem harsh and restrictive but on the other hand forces authors to actually include details on these subtopics. The quality of the language and style and the structure of the paper are two important factors that determine the overall clarity of a paper. In other words, how easy is it for a future reader to follow the train of thoughts of the authors? A good reviewer assesses whether a paper is readable only by an in-crowd of specialists or whether a less specialized reader generally knowledgeable of the overall domain can understand the text. Certain journal editors ask reviewers explicitly to judge a paper on the breadth of its potential target group of readers.

  Even if a paper structure is not explicitly imposed, it remains a good "memory aid" for a reviewer to assess whether or not the most important aspects of a scientific paper are dealt with by the submitting authors or at least are explained why these aspects are not included. For example, a lacking evaluation can be justified as the work described is still in the early stage,[9] but at least it must be mentioned. Authors have to anticipate on questions readers could raise. Reviewers have to check whether authors succeeded in adequately replying on potential questions by readers.

- Formulas/Formalization

  Computer science papers easily contain mathematical formulas or formalizations in general. Usually their purpose is to avoid ambiguity in definitions and prove soundness of an algorithm and/or the time and space complexity of the proposed solution. However, formalizations as such are empty when not defined and explained properly. Authors cannot expect readers to best guess what the exact meaning of the symbols are (even if some symbols are used very often with the same meaning). It is the task of reviewers to check if formal definitions are available, consistently used, and correct and do not contradict what is written elsewhere in the paper in natural language. Using formalization in a paper has to add to (the clarity of) the discussion. It should not merely add a cosmetic flavor of seriousness to a paper or to distract the attention from some reasoning flaw.

- Tables/Data

  It is very common to find tables or some other way to schematically organize data in papers, in particular in discussion or evaluation sections. Reviewers should not simply take for granted the interpretations or explanations authors attribute to data. It happens regularly that authors overgeneralize trends, "stretch" how they interpret data to support their theory, or simply draw a wrong conclusion. It is good reviewer practice to scrupulously verify that statements made by authors are really justified by the data.

  Structuring the data in a visually supportive way to guide a reader toward the statement(s) or conclusion(s) proposed by authors can be helpful but at the same time misleading. Reviewers have to ponder whether the data presented is complete. For example, would the study of another parameter lead to other data falsifying the hypotheses put forward by the authors? Would a reorganization of the data shed a new light on the matter researched? It is possible that these kinds of reflections point to weaknesses or overlooked angles of a (complex) research question. It is the job of a reviewer to consider such issues.

[9] This situation is more likely to be the case for workshop submissions. A journal submission, which as a rule discusses a substantial part of work done, cannot do without.

- Graphics (Layout)

  Napoleon said that a drawing is worth a thousand words. Graphics not only visually "lighten up" a paper but allow for the succinct representation of complex and complicated statements that otherwise would be difficult to describe in natural language (unless in a very lengthy and verbose way). Consequently, graphics are very important tools to pass on information. But again, they might lead a reader on the wrong path. For example, the dimensions might suggest that a trend is more outspoken than it actually is. Or on the contrary, using a logarithmic scale visually hides a strong increasing evolution. Reviewers are supposed to verify that the type of graphic (e.g., pie chart vs. scatter plot) is the best one suited, that legends are adequate and self-explanatory, that axes are appropriately defined, that accompanying text accurately describes certain choices (e.g., removal of outliers), and that captions fit what is visually represented in a self-containing way.

  Usually specific guidelines and rules on how to format a graphic are imposed on authors. For example, captions are to be placed below the graphic. Another frequently occurring problem is the readability of a graphic (too many things cramped in a too limited space). Or sometimes, authors forget that color graphics can become "ambiguous" or unclear when printed in gray scale. Although quite often (mostly for journals), publishers do these kinds of checks themselves, reviewers should warn authors about problems with graphics in view of the preparation of a good camera-ready version.
- Statistics

  Authors often make use of statistics to prove the validity of results. For example, user studies or evaluations involving a reference or gold standard usually calculate whether or not results (e.g., similar behavior of two setups) are due to chance and which level of significance the results have. Statistics, as formalizations, increase the perception of solidity and seriousness of the work, allowing for the generalization of the results. However, depending on the exact setup of the experiments, different statistical tests apply. Reviewers should therefore examine if the statistical tests that the authors apply are indeed the correct ones and whether or not the authors correctly accept or reject the initial hypotheses.
- Innovativeness

  Innovativeness or originality of the work is a popular criterion in paper (and proposal) reviews. On the one hand, this can create an inflation of so-called news claims, buzzwords, or techniques without existing ones being properly or exhaustively researched or verified/falsified. On the other hand, submitted papers should contain a sufficient amount of new insights to warrant a publication. Conference chairs or journal editors usually include in the review instructions how strong the focus on innovativeness is.

2.2.2 *The Review Subject (The Reviewer Himself/Herself)*

Not only the aspects of a paper under review are important but also how a reviewer articulates his/her comments, suggestions, and critiques. In a sense, review writing can equally be considered as a skill a junior researcher needs to develop. The main guiding principle can be summarized by the question "would you react angrily when receiving your own (negative) review?" In the following sections we present some aspects of a "reviewing etiquette."

- Be Helpful

  A review is in essence a kind of evaluation (usually with explicit scores and a final judgment: accept or reject). However, a good evaluation not only consists of a judgment but also of suggestions for improvement. Consequently, good reviews are never limited to scores, a few lines of comments, and a final acceptance/rejection decision. On the contrary, good reviews contain constructive critique and suggestions for methodological and textual improvements and point to missing references, lacking details, flaws, or paragraphs difficult to understand.

  Even if receiving a notification of rejection, an author, after his/her initial disappointment, should be satisfied with the constructive comments. Unfortunately, many reviewers, in particular of workshops, do not adopt a very helpful attitude and are quite succinct (or even plainly blunt) in their comments (cf. Sect. 3.1.12). Submitting authors, in particular junior ones, have a right to know what the problems with their paper are and want to be assured that their paper has been reviewed in a serious way. Elaborated and respectful comments (though taking up quite some time of the reviewer[10]) should be standard. Extensive comments "prove" to the submitting authors that reviewers spent sufficient time to thoroughly read the submitted paper, that they understand what the authors

[10] Some conferences are awarding a prize for the best reviewer as otherwise the effort put into reviewing remains unnoticed in the academic track record.

are conveying (cf. Sect. 3.2.2), and that they have adequately explicated their arguments in favor or against.[11]

- Remain Factual

  Science is a matter of facts sustaining or falsifying hypotheses and of finding solidly grounded answers to research questions. Although reviewers are not emotionless robots, as a reviewer you should avoid using emotional wording or other words indicating an intuition (e.g., "feel") or speculation (e.g., "suspect"). Arguments (pros or contras) are preferably expressed in a factual manner based on the paper content or available literature. Whether a reviewer "likes" or "dislikes" a paper is irrelevant. The basic question a reviewer has to answer is whether the work presented in a paper was performed according to scientific standards and criteria. Additionally, the reviewer has to check if the proposed work is important enough to be included in the proceedings or journal. Finally, the readability of the submitted text has to be evaluated.

  A good trick to write factual reviews is to explain why you, as a reviewer, have attributed a certain score to an evaluation criterion (cf. Sect. 3.1.6). For example, why is a paper less or not innovative: Is the research question already (conclusively?) dealt with by other (which?) researchers? Are the authors simply repackaging too much previous work? In particular when the review form consists of a single free entry text field, a reviewer must be disciplined enough to address all the criteria instead of going into some form of "free flow mode." Especially if a review is negative and/or suggests that the paper is to be rejected, it is important to link the review comments to the criteria (and scores). Otherwise, authors might get the impression that a reviewer did not follow the review instructions properly and came up with some half cooked, hastily improvised impression.

- Check Your Language and Style Quality

  Not only does the submitting author have to pay attention to a correct use of the language and style, this holds for reviewers as well. After all, if authors are not able to understand your comments, there was no point in providing them with comments. Therefore, it is good practice to reread your review text and specifically check the spelling, language, style, and understandability. Authors usually do not have a high esteem of reviewers who make comments on language errors but do so in a form of "coal miners English." Nowadays browsers have built-in spell-check

[11] Elaborated comments surely deserve an acknowledgement or other explicit mentions by the authors.

capabilities (when you fill out a review form online). Another trick is to use your favorite word processor with spell and style checking tools to write your review and to copy/paste the text afterward into the review form text boxes. In addition, to facilitate life for a revising author, always clearly indicate the precise location (page, section, paragraph, line) in the paper to which your comments apply.[12]

- Take the Specific Context into Account

  Conferences, workshops, special journal issues, and journals in general all have their specific thematic focus, target community, geographic span, sometimes language area, review criteria and procedures and infrastructure, time lines, and formatting guidelines, which are detailed in the call for papers. A good reviewer has to make himself/herself acquainted with these idiosyncrasies and comply with them. For example, it is pointless to reject a submission due to a lacking evaluation section if the call for contributions targets short position papers with visions for future trends. It is bad for your reputation as a reviewer when chairs and submitters discover that you did not bother to familiarize yourself with the context of the call for papers.

  It also helps to estimate the time and effort you may want to spend on reviewing. Reviewing for a high-impact journal requires more time, attention, and scrutiny than an informally organized workshop. Of course, this is not a plea minimizing on time spent on reviews for workshops.

- Dare to Decide

  Organizers or editors invite reviewers for their expertise and renown and entrust them with the responsibility to help them to decide on acceptance or rejection. A neutral position is thus in general not very useful. In case of a neutral review opinion, at least add some explanations (or guiding comments) (possibly in the text field for the chairs' or editors' eyes only). In general, do not hesitate to reject. "Fresh" reviewers seem to show some more lenient tendency toward acceptance, while very experienced reviewers on the other hand can be more inclined to reject more easily. A useful trick is to give the scores on the different criteria (together with detailed descriptions of all the issues, positive and negative, that you have identified in the submission) and formulate your questions to the authors first, before making up your mind on global acceptance or rejection at the end after having reexamined your partial scores and comments. It implies that it is not a good practice to read a

[12] Some journals apply a line numbering format for submissions.

paper only once, immediately decide to accept or reject it, and subsequently write your comments and provide partial scores toward your initial overall opinion. Late reviewers show such behavior but it is not very respectful and sometimes not even helpful (cf. above). Good reviewers read the paper several times with different "mind sets," e.g., a first time to get acquainted with the paper in general, a second time with a critical mind to produce a first version of the review, and a third time with a focus on specific parts of the paper when finalizing the review.

2.3 To Conclude

This chapter presented an overview of the most important principles and attitudes a good reviewer should adhere to. In addition, some relevant practical aspects have been discussed of what a good reviewing behavior consists of. This not only presupposes a certain personal attitude of a reviewer but also concerns specific elements of a scientific paper that a good reviewer should look at in detail.

Chapter 3
Tips and Tricks

3.1 Good Reviewers Keep the Following in Mind

3.1.1 Do Not Waste Your Time on Sloppy Authors

Although "accidents" (e.g., submission of the wrong paper, unclear call for papers, or wrong style sheet) can always occur, some submissions are clearly out of scope and unfit for a specific conference or workshop. These papers have only a weak connection to the specific conference themes or clearly do not follow the specific guidelines of a call for papers. Obviously, these authors did not bother to "tune" (both content and form) their paper to the conference or workshop. Some even forget to replace the style sheet that reveals in a footnote the name of another conference. Such authors might also be searching for an easy way to receive external comments on their research work without considering to actually attend the conference.[1] Or, they simply resubmit the same paper that has already been submitted and possibly rejected elsewhere.

Although a reviewer should not conclude too quickly that a paper is "out of scope" and therefore simply dismiss it, on the other hand, a reviewer is not an easily available external mentor for free either. An extreme example is a submission of a content-wise largely out-of-scope "paper" of more than 45 pages, although the call for papers clearly mentioned a maximum of

[1] Another, equally reprehensible, reason is that, e.g., workshop organizers stimulate people to submit "something" that will be rejected anyhow but that helps to obtain a high rejection rate. A high rejection rate is reputation-wise interesting or simply imposed by a publisher and/or conference steering committee.

P. Spyns, M.-E. Vidal, *Scientific Peer Reviewing*,
DOI 10.1007/978-3-319-25084-7_3

15 pages. Do not hesitate to reject such an out-of-scope paper without much ado. A reviewer can use his/her time more wisely than writing a serious review for a submission that very obviously does not respect the guidelines of the call for papers. If however the authors did a genuinely serious attempt in writing a paper but failed for some reason (e.g., first paper by junior researchers, undetected methodological flaw, or extremely bad paper structure), you are indeed supposed to clearly explain the main problem(s) (and provide some examples) but not to comment on all the ensuing problems or flaws.

3.1.2 Avoid Repetitions

Be careful not to use the same sentences, expressions, or thoughts too often in a row, in particular if your review is not that long.

Example 1: (entire review of a paper submitted to an international workshop)

Reviewer1: [overall: accept]

Summary: *The paper explains in detail the X methodologies and processes.*

Details: *The paper explains process of X in detail. It also compares existing work in the area and clearly identifies the future work need. The paper can be accepted for Y 2010.*

Reviewer2: [overall: weak accept]

Summary: *The author evaluated the quality of X methods, which is important for the long-term development and evolution of Z.*

Details: *This paper explains the method and criteria used to evaluate the quality of X approaches.*

However, this paper is lacking in detailed description of the results of the experiments. The use of a diagram or table would have made the key findings of the paper more legible.

Both reviewers simply rephrase in the "details part" what they have written in the "summary part," which already is a very summary description of the paper content. Furthermore, reviewer 1 simply describes the paper structure without any further assessment of the "value" of the paper content. At least, reviewer 2 is a bit more critical, although the comments remain very general. It is evident that these two reviewers did not behave in a serious and professional way. Even though the result was positive (paper accepted), the author felt disappointed to receive such a superficial review of his (hard) work as the comments hardly or do not address specific points to improve the paper.

3.1.3 Never Include Details of a Personal Nature

Reviews are factual and neutral assessments of scientific work. Although reviewers surely can include encouraging statements or expressions showing their (dis)approval, too personal details are not appropriate. In particular, indicate clearly whether your statements or comments are backed by data (in the paper) or existing literature or when they are personal opinions.

Example 2: (review of a paper submitted to an international workshop)
Reviewer4: [overall: weak reject]
Due to my vacation the review is in staccato.
I reject the paper I think it doesn't add something new, if you use X, Y is standard regardless the domain. In this case the domain is Z.
I agree that Y improves the quality and is a much more natural way to model Z. But in this case I would like to see the Q completed.

Most probably no author (as well as no review committee chair nor workshop organizer) is happy to learn that his/her paper is reviewed by someone who explicitly indicates being in a holiday mood. Nevertheless, the reviewer immediately adds that he/she did not discover anything new in this paper, without any reference to the literature to corroborate this opinion. As two other reviewers were (much) more positive and provided more elaborated comments, the author complained to the workshop organizer who promised not to invite this reviewer any more.[2]

Reviewers should be very careful when using "strong" comments. They should at all times behave in a respectful manner and adopt an attitude that shows courtesy. Authors and reviewers alike might not master all the subtleties of the English language and could misinterpret or misread the level of emphasis or "strength" of certain expressions.

Example 3: (excerpt of detailed comments on a paper submitted to a high profile international conference)
Reviewer3: [overall: weak reject]
The paper is rather pedantic, and longer than necessary. [...] I would have preferred a paper defining "pragmatic" as "practical, realistic, non-nonsense, down-to-earth" :-(

"Pedantic" has a clearly negative connotation (intentionally using overly difficult phrasings to hide one's lack of knowledge or to give an inflated perception of one's actually limited knowledge or importance). By this

[2] Due to the low score of this unprofessional reviewer, the paper was the first paper in rank that failed to pass the acceptance threshold. However, an improved version was accepted the subsequent year.

specific word, combined with the other comment and the negative emoticon, the reviewer blames the authors of showing off and being insufficiently knowledgeable. Without explicit and specific substantiation (e.g., references to related literature) of such a negative opinion, reviewers should take care to avoid such statements.

Example 4: (excerpt of detailed comments on a paper submitted to a high profile international conference)

Reviewer2: [overall: reject]

Instead, the authors extensively refer to philosophical literature that does indeed discuss the general problem but has little to say about technology suitable to overcome them. As a result of this bias, the motivation for the work is quite convincing but the description of the proposed solution is neither convincing nor even described in a suitable way. (...) Even if we neglect formal aspects of representing and reasoning about different viewpoints, I have a problem with the technical part of the paper, as it does not enable me to understand how the proposed K M is used in communication, i.e. *in which way it behaves different from standard Ms. (...) As I mainly see the SW track as a technical track that should discuss new technologies, I have to vote for rejection. However, I would like to encourage the authors to elaborate on the technical aspects of the D approach and re-submit the paper to an event that is explicitly dedicated to the problem of C or M N (*e.g. *"C 200x" or the "D'0x Workshop on Xs and D Systems"*

Although this reviewer also rejects the submission (which is the same as in Example 3), for being too theoretical, the wording is completely different: helpful (suggesting another conference), stimulating (acknowledging the importance of the problem), well elaborated (pointing to specific weak points), and respectful. Even if the net result is worse (reject vs. weak reject), the perception by the author is completely different and much less negative.

3.1.4 Elaborate Your One/Two-Liners

Reviews—even positive ones—have to be substantiated. It is very rare that submissions cannot be improved in any sense or do not generate any questions or reflections on the part of the reviewers.

Example 5: (all detailed comments on a paper submitted to an international workshop)

Reviewer1: [5-Excellent paper]

Comments: *Please consider applying it to or using a business case.*

Reviewer2: [4-Should be included]

Comments: *Very good work. Keep it up.*

Reviewer3: [4-Should be included]
Comments: *Accept—no changes required*

Reviews like these are simply too positive to be true. It almost looks like "pro forma" reviews performed by the workshop organizers themselves to make sure a submission is accepted. A good reviewer never uses predetermined "templates" as this for reviewing.

Example 6: (excerpt of a review for a paper submitted to an international workshop)

Reviewer: [overall: accept]

- Validation: Does the paper show convincing evidence to support the approach? Do the authors present solid arguments about their position?

  Yes.
- Current status: Does the paper state clearly how much work has been done to date, i.e., what has been designed, what has been implemented, and what has been evaluated?

  Yes.

Although the review form contains a list of specific review questions that address various aspects of good scientific paper writing and that can be answered by a simple "yes" or "no," it is nevertheless a good practice to provide a more elaborated answer.

Example 7: (entire review of a paper submitted to an international conference)

Reviewer2:
Originality: *Accept*
Quality: *Accept*
Relevance: *Accept*
Presentation: *Strong Accept*
Recommendation: *Accept*

Summary: *The author presents experiment results of an automated X evaluation procedure. The experiments are to test if the automated procedure can detect the quality of an Y output.*

Details: *A timely experiment in the critical area of IS interoperability. The paper is well-presented in a logical manner with substantial research value and supporting experimental sequence(s).*

This very positive reviewer has shortly explained why he/she likes the submitted paper but apparently sees no further issues for improvements nor supplies any suggestion for future work. Such kind of reviews may at first sight be welcomed by the submitting authors but do not provide much extra directions for future research.

Example 8: (excerpts from reviewer detailed comments on a paper submitted to an international conference)

Reviewer4:
Originality: *Weak Accept*
Quality: *Neutral*
Relevance: *Accept*
Presentation: *Accept*
Recommendation: *Weak Accept*

Details: *This is an important problem because there are several methods for building an adaptive X and there are several manual or automatic preprocessing mechanisms that can be applied to the data in order to facilitate the discovery of an X. (...) I liked the paper. The experimental setup seems well conceived, but the results are not conclusive. However, it would make a nice experience paper on a topic that is very important but also not very well understood.*

Although the results have not been conclusive, it is not clear whether the reason is the properties of the specific dataset or more general reason on the quality of the E P itself. So perhaps another way to prove the paper would be to make more comprehensive experimental evaluation.

This reviewer also likes the submission. But most importantly he/she provides suggestions to the author to improve the paper. Even if you as a reviewer are positive about a paper, you should shortly explain your reasons for being positive and nevertheless point out some possibilities for improvement and/or suggestions for future work.

3.1.5 Check on Student Reviews

Although in practice it often happens that Ph.D. students do review work for their supervisor or head of the research lab, the latter should not forget to review their reviews as he/she still remains the main responsible for the review. In particular, if a Ph.D. student is less experienced and is still learning to review "by doing," it is not only disrespectful toward the submitting authors not to check on the Ph.D. student's work but also fallacious toward the program chairs (as they did not invite the Ph.D. student as a reviewer).

Example 9: three complete reviews (detailed comments only) by three Ph.D. students of the same supervisor[3] on different submissions to an international Ph. D. workshop

Paper 1: [overall recommendation: weak accept]
Comments: *The paper is well written and organized.*

---

[3] In all fairness, the Ph.D. students did also include extensive summaries of each paper, but the summary was only a description, not an assessment, and hence useless as a review.

Paper 2: [overall recommendation: weak reject]
Comments:
The article is well written and organized. However, it misses details in the state of the art analysis:

- *More details on the used categories (definitions, . . .)*
- *The application domains of the approaches*
- *Details on the most relevant approaches.*

Paper 3: [overall recommendation: neutral]
Comments:
The paper is in general well written. It misses however some clarification that would help the reader to understand the approach.

None of the Ph.D. students gives a comment that addresses the paper content in depth. In addition, the comments are very general and reusable. In fact, all reviews lack "some clarification that would help the authors to understand the comments." As even by Ph.D. students' standards, these reviews are utterly bad, it really is a blatant shame that their supervisor apparently did not bother to redress the situation (lack of time? lack of interest? lack of professionalism?). This behavior is deadly for the scientific credibility of reviewers (and even of the Ph.D. students). Needless to say this reviewer was no longer invited by the program chairs for the next issue of the workshop (and forever blacklisted).

3.1.6 Structure Your Review Comments According to the Evaluation Criteria and Make Them Correspond to Your Marks (If Applicable)

Scores and comments are not independent items in a review. A score is a numerical summary of a reviewer's opinion but has to be related to the freestyle "detailed comments to the author." Otherwise, the submitting author cannot really improve his/her paper not knowing exactly what the problem is (only that there is some problem).

Example 10: (entire review of a paper submitted to an international conference)

Reviewers Familiarity: *6*
Evaluation of work and contribution: *7*
Significance to theory and practice: *6*
Originality and novelty: *6*
Relevance: *7*
Readability and organization: *7*

Overall recommendation: 7
Summary:
A formalization of X together with Y is proposed. Approach described, no tests so far. Work-in-progress.
Comments to the author:

In the review above, no specific comments for improvement are provided. Apparently the paper is judged to be not that novel or original, but the reviewer does not provide any specific comments at all, e.g., references to the scientific literature to prove his/her point. Reviews should be more than numerical summaries of personal opinions that are of little help to the submitting authors.

3.1.7 Always Mention Some Positive Aspects (Strengths) of the Paper as well as Some Weak Points as Suggestions to Improve the Quality of the Paper

Papers seldom are that good (or bad) that no weak (or strong) points can be suggested to the submitting author(s). Mentioning strong and weak points shows that a reviewer took his/her job seriously and reflected on all aspects of a paper.

Example 11: ("for author" field comments on a paper submitted to a high profile international conference)
Reviewer3: [reject]
For Author: *This research direction is promising, but more effort needs to go into this project. Investigate more carefully the effects of different X algorithms, parameter settings of these algorithms, methods for extracting Y from Z, etc. This should be reflected in a rewritten paper that allocates significantly more space to the utilized algorithms and performed experiments while condensing the overview and eliminating "boiler plate" paragraphs.*

Though the reviewer has rejected the paper, he/she expresses a positive opinion on the research direction and provides helpful comments on which aspects should be improved. The submitting author is now able to understand why his/her paper was rejected and might be well motivated to improve the weaknesses mentioned.

Example 12: (complete reviewer detailed comments on a submission to an international Ph.D. workshop)
Reviewer1: [overall recommendation: weak accept]
The paper has some strengths and weaknesses: strengths:

1. *It is a valuable idea to handle heterogeneity in the approaches.*
2. *It provides proof of processing tasks with challenging complexity and quantity.*
3. *Interesting idea to apply semantic web knowledge.*

Weaknesses:

1. *Not be able to provide data access.*
2. *The current approach is not substantiated.*
3. *Even though it uses semantic web, it doesn't give enough details about how they used.*

Other reviewers organize their comments as a list of strong vs. weak points, usually followed by a conclusion with the overall opinion of the reviewer on the paper. The review above is well articulated. The comments however are too succinct and not specific enough (probably as they were given by a junior Ph.D. student).

Example 13: (detailed comments on a paper submitted to an international symposium)

Reviewer4:
Please provide constructive comments to help author improve the paper.
RECOMMENDATION: *Do not accept.*

Although the review form explicitly asks for constructive comments, the reviewer blatantly rejects the paper without any explanation at all. In particular, when a paper is rejected, reviewers should include sufficient detailed comments that explain the main limitations and drawbacks of the work. If no comment whatsoever is given, authors might feel that other, nonscientific factors have played a role, which is unacceptable. For example, a reviewer might want to stall the publication as he/she is working on the same topic and wants to be the first to publish. In general, providing no comments is not helpful at all, is very disrespectful, and, hence, cannot be considered as good reviewing practice.

Also authors may doubt that the reviewer did actually read the entire paper or understood their work. Chairs or editors should monitor incoming reviews and dare to ask reviewers for more extensive and elaborated comments—as in the case of the example above—even if it concerns a high-profile researcher. The good reputation of conferences or workshops might suffer seriously from too many nonmotivated rejections.

3.1.8 Include Specific Questions to the Authors

It is good to engage in some kind of conversation with the authors by including specific questions. Questions demonstrate that a reviewer "thinks along" with the author. They can suggest or lead to methodological improvements, textual clarifications, alternative experiments, increased algorithmic efficiency, additional conclusions, research avenues for future work, etc.

Example 14: (excerpt from reviewer detailed comments on a submission to an international Ph.D. workshop)

Reviewer3:

Approach seems interesting with a preliminary result. These results however lack from precise context and behavior detected: what are real and precise examples of context—without this level of detail, it's hard to check if the classification is appropriate or is classifying totally nonrelated situations? Is the data mining resilient to context changes? False positive/negative (especially when emergency cases are involved)? What can be optimized with the result of this classification? What kind of situations benefit from it?

This reviewer (a Ph.D. student!) clearly made a serious and appreciable effort by asking relevant questions to the authors. Nevertheless, as a reviewer you must also ponder on what you can realistically and fairly expect an author to rework or revise. For example, do not ask an author to add a new experiment and rewrite half of his/her paper if the camera-ready version is due within a week. Be clear in what an author absolutely has to adapt for the next paper version to be acceptable and what he/she could/should do in the long run in his/her research activities.

Example 15: (complete reviewer detailed comments on a paper submitted to an international conference)

Reviewer2:

Originality: *Accept*
Quality: *Neutral*
Relevance: *Accept*
Presentation: *Neutral*
Recommendation: *Neutral*

Details: *I like papers that show how certain approaches don't work, and we should have a lot of such papers that give honest views of the results, and that extend their evaluations to check for other anomalies.*

However, the reason why I've given this paper a Neutral recommendation is because it seems to overlook the "why" and focus too much on the "what."

*For example, it is interesting to find out that removing T did not improve things at all (or made them worse), but *why* was this the case?! If we don't know why then I think it's difficult to draw any conclusions from this behavior.*

Was it because they did not remove enough T? Was it that the T they removed (4443 W) was a repetitive T? How many unique Ws were in those 4443 ones? Did they remove repetitive H and N and such things? Or did they remove standard T? What happened when that T was removed? Why did incorrect Ys start to crop up and why were they not present in the first run? I think without actually investigating this more deeply, we will not be able to appreciate the full picture of this event. My conclusion is that your approach did not work with the full T, and it also did not work with the cropped T, so I would question the whole approach and not the T itself because clearly the T has little effect here!

Why did you get more unique Ys in the second experiment? I would have thought that you should only get less Ys, not so many new ones. Can you explain this please? With an example?

The authors conclude that perhaps Q is not sophisticated enough which is why they are witnessing such behavior. Q is not detailed in this paper as it was already published elsewhere (and cited here). Doubting it is a valid point of course, but this questions the results of this paper even more, since now we don't really know whether the fault lies with Q, or whether removing T generally does not make any difference. So once again, the "why" is not well answered, and the readers are more puzzled.

One last point, how different would the case be if you use some other X L tools? How biased are these results to your choice of extracting Ys? I would guess that other tools probably use different approaches that are not necessarily purely limited to statistical analysis.

Overall, a nice paper, but lacks enough scientific depth to make its conclusions stand.

In the example above, the reviewer clearly explains in a neutral way what is lacking in the paper by elaborately substantiating his/her overall judgment with many specific and to-the-point questions to the author on the design of his/her experiment. As probably other readers would have similar questions, the author received valuable suggestions to make the subsequent version of the paper more scientifically attractive and interesting.

3.1.9 Check the List of References

Different scientific domains use a different style for the references (usually included in the formatting guidelines). Reviewers should not forget to check if the references are appropriate, up to date, and complete (many authors do not mention page numbers anymore or expect that readers are familiar with abbreviated conference or workshop names). Although many large commercial publishers nowadays have a dedicated team of correctors who check the formal details and completeness of bibliographic references, it remains the task of a reviewer to also examine the list of references.

Example 16: (excerpt from detailed comments on a submission to an international Ph.D. workshop)

Reviewer2:

- *Check the citation list. Use English for geographic locations. If a source is not written in English, then of course use the original title, mention which language it is in, but still include the geographic information in English (Germany and not Alemania, Spain and not Espana, etc.).*

This reviewer (a Ph.D. student!) very correctly points out what needs to be done when compiling the reference list.

Some reviewers cannot resist the temptation to suggest in their review comments that citations to their own work or publications are to be added to the bibliography section. Especially when a reviewer only promotes her/his personal work, this practice can be interpreted as a dishonest way of self-promotion for mere citation purposes. All too often authors, as they have to take the review comments and suggestions into account when producing a revision or camera-ready version, simply include the references suggested to avoid another round of revisions. Although such suggestions may be very valid and justified, it remains a delicate matter. All references suggested must be directly relevant for the work of the paper.

A reviewer must be very cautious in giving a bad score or negative comments for the bibliography or related work section only because his/her papers are not cited. This practice can certainly qualify as misconduct as the reviewer has the "power" to reject the paper and is protected by anonymity—even though he/she takes a risk of being "recognized" by the specificity of his/her references. Only when his/her work clearly is an important reference in the domain and highly relevant for a certain topic, he/she may penalize their absence.

Example 17: (excerpts of reviewer's detailed comments on papers submitted to international conferences)

Due to the topic of the work I highly recommend the authors to have a look at G and S: S M, 2003/04, who have presented some very similar ideas in their work.

Finally, the authors might want to check the work of H and his group on C-based S M at the U of W.

Some additional relevant references concerning modelling/conceptualization are the following: ...

It is generally best to propose additional bibliographic references as neutral suggestions or gentle recommendations and give the author(s) the liberty of deciding on the relevance or appropriateness of the suggestions.[4]

[4] In some cases, mainly journal paper reviews, authors are expected to explain in a separate letter to the editor (and reviewers) how they have taken the comments by the reviewers into account. They can thus argue about the validity of the references suggested by reviewers and defend themselves for not including the suggested references anyhow.

3.1.10 Search the Internet Using the Title and/or Abstract

Often, in particular at conferences, a paper describes work at some stage of the research process. Previous work or preliminary reflections may already be published so that the current paper under examination builds on earlier papers. Some degree of repetition of information is thus unavoidable. Journal papers summarize the entire research process on a certain topic and inevitably "repackage" earlier publications. Nevertheless, one cannot simply copy and paste large chunks of text from earlier publications into newer ones ("text recycling") or glue together several conference papers into a journal paper. And in some parts of the world, even copying and pasting text from papers by others without referencing is not always unequivocally recognized as academic misconduct.

A good reviewer is sufficiently aware of the existing literature to signal to the chair or editor (e.g., using the "comments to the chair/editor" form field) if submitted papers contain too much material already published. The ACM and IEEE have guidelines on including conference contributions in journal papers.[5] Usually the chair examines the matter and takes an ad hoc decision depending on the nature of the publication channel[6] and the amount of duplicate information.[7] Workshop organizers might be more lenient in order to fill up the program and stimulate discussions.

A simple and effective trick is to copy/paste the title and/or abstract in an Internet search engine and see if similar (nonreferenced) publications by the same author(s) pop up. If you, as a reviewer, think the matter is too delicate to decide for yourself,[8] whether or not to reject a submission for this reason, you should mention this in the "comments only for the editors/chairs" section of the review form. For example, it was very easy to discover that a journal paper was not much more than a concatenation of two conference papers (a technique also called "salami slicing") or that a conference paper was a slightly extended workshop paper of the year before. But do not be

---

[5] Also see the COPE guidelines: http://publicationethics.org/text-recycling-guidelines

[6] For example, a technical report on the researcher's website or a workshop publication on the CEUR site is usually not considered as normal publications.

[7] How many paragraphs, and in which sections, have been copied/pasted? Are many sentences literally repeated? What is the proportion of new vs. old information? Etc.

[8] Copy/paste repetitions in introductory sections could "enjoy" a more permissive attitude than repetitions in the discussion or result sections. In some non-Western cultures (self) plagiarism seems to be more frequently applied and apparently treated more forgivingly (which nevertheless cannot be accepted!).

surprised if, e.g., workshop chairs do include this paper—as they want to fill up their program. Using a search engine might also prove useful regarding the related work section of a submitted paper, as nowadays it is difficult to keep up with all forms of scientific literature, even in a limited domain.

Reviewers should also be able to detect obvious forms of plagiarism. As the body of scientific literature is continuously and exponentially expanding, it has become increasingly difficult to keep up with all (important) new publications (even in smaller domains). Therefore, journal editors (and some conference organizers) more and more use software such as CrossCheck or iThenticate to detect plagiarism.[9]

3.1.11 Differentiate Your Reviews

Although in essence, the reviewing process is more or less the same irrespective of the type of "target" submitted to, in practice there is a difference between a review for a journal, conference, or workshop. The former takes longer and is more demanding, while the latter can be dealt with in a relatively quicker manner. Nevertheless, a review for a workshop still has to meet the abovementioned qualitative thresholds. But it is evident that for a journal paper, as it represents a larger chunk of work, more comments are requested and more scrutiny is needed. Also a journal usually has a high profile (impact, editorial board, etc.) it wants to uphold by being highly selective, hence requiring well-motivated and substantiated review comments. A workshop has a more informal character: the contributions usually concern early work in progress so the reviewing process usually is not that "heavy" compared to a journal. Conferences are somewhat in between. As a reviewer you should spend your reviewing time accordingly. Although it is generous of a reviewer to give more than a page of comments on a workshop paper (and certainly well appreciated by the authors and chairs), there are too many workshops around to maintain such a reviewing time investment. Conversely, make sure to well prepare a journal review, as invitations to review for a journal indicate your academic status in the field (and hence are interesting items on your track record). Bad reviewers are usually invited only once.

[9] Cf. https://en.wikipedia.org/wiki/Plagiarism_detection

3.1.12 Give Accurate and Specific Comments

Comments, if given, are only useful to the extent that the authors can understand how they have to address them. Good comments clearly identify the issue (also by unequivocally identifying the location in the paper of the problem) and suggest ways to remediate or fix the problem.

Example 18: (complete reviews of a paper submitted to a high profile international conference)

Reviewer2:
Please rate on a scale from 1 to 10 (1—very poor, 10—excellent)
Relevance to the workshop: _*8*__
Importance of the problem: _*4*__
Originality and novelty of the solution: _*2*__
Technical quality: _*2*__
Readability and organization: _*5*__
Please summarize the paper:
The paper is providing a refinement of the notion of Y in the X framework.
Comments for the authors:
I don't see any new aspects in the paper. The paper is just summarizing what exists.

Reviewer3:
Please rate on a scale from 1 to 10 (1—very poor, 10—excellent)
Relevance to the workshop: _*7*__
Importance of the problem: _*8*__
Originality and novelty of the solution: _*2*__
Technical quality: _*2*__
Readability and organization: _*4*__
Please summarize the paper:
This paper attempts to cover the subject of P to Q within X framework for Y.
Comments for the authors:
This is a very important subject, but the treatment of the subject appears to be rather incomplete and not novel.

Although the two reviewers express the same opinion (the paper lacks novelty), none of them provides references to what exists (which could also be a comment on the related work section). In addition reviewer 3 judges the paper as incomplete without mentioning what should be included in his/her view. Of course, it remains the responsibility of the author in the first place to assure "a complete treatment of the subject." But such a general comment is all too easy and fast to write down, can apply to many papers, and can be written down by anybody (seasoned researcher or junior Ph.D. student), leaving the author to guess what should be included (might be details?) and is simply unworthy of a high-profile international conference. The editor or chair can only rely on the reputation of the reviewer to appreciate the value of such comments and scores. Having too many papers to review and not enough time simply cannot be accepted as excuses from the reviewers to deliver such a (negative) one-liner.

Example 19: (complete reviewer detailed comments on a paper submitted to an international conference)

Reviewer4:
Originality: *Weak Reject*
Quality: *Neutral*
Relevance: *Strong Accept*
Presentation: *Accept*
Recommendation: *Weak Reject*

Details: *I liked portions of this work and would like to encourage the authors to work along those lines.*

The distinction between a single ES and multiple ESs is superfluous and also the analysis by P never made this assumption. The authors should better relate their characterization of ESs with the one presented by P.

The authors have looked at the issue of the formal/nonformal semantics and issues related to adoption... individual vs community focused. One glaring omission in their analysis is the fact that we need to formalize a P for generating consensus (whether based on usage or expert opinion) to move it from an F to an X... There is work in cultural anthropology and group communication theory which is relevant to this activity... I would encourage the authors to look at the following papers: (...)

There is an interesting paper A on weak and strong semantics you should look at.

The creation of a new Q prototype is important and interesting, but doesn't add anything to the research in the area.

Although the reviewer of the example above likes the overall idea expressed in the submission, he/she does not hesitate to point the author to some specific weaknesses (a misinterpretation of related work, a missing related work in another domain, a missing reference, an absent formalization, and a superfluous prototype). In addition, he/she gives specific bibliographic references the authors should study. The submitting authors have received sufficiently concrete comments and suggestions for future research that will strengthen their results.

3.1.13 Review Independently

Some online reviewing systems allow reviewers to see the comments and scores of their peer reviewers. Usually this is only possible after a reviewer has submitted his/her review and scores. Obviously you should input your own review and scores first without being influenced by how other reviewers think about the paper. Some lazy, late, incompetent, or unsecure reviewers read the comments of their peers first and then rephrase these when writing their comments.[10] Without any doubt this is bad practice and even reviewer misconduct!

Of course, most reviewers are curious to see to what degree their opinion concurs with the one of the other reviewers. It may happen that some reviewers adapt their scores and comments afterward. As long as this does not comprise a complete overhaul or drastic change, this might even be helpful for a chair or editor as usually this results in more converging opinions. But be aware that many reviewing systems nowadays log every change a reviewer makes. Chairs or editors are able to detect if a reviewer tries to conform or overly align his/her scores and comments with those of other reviewers. Even though chairs or editors usually prefer opinions and scores that do not diverge too much, they are not interested in receiving "recycled" or all too similar reviews either. In short, make up your own mind (and review) first and stick more or less to it.

3.2 Some Miscellaneous Issues

3.2.1 How to Review a Submission by (Students of) the Chair or Organizer?

A peculiar situation arises when a reviewer is asked to review a paper submitted by a chair or workshop organizer (or his/her students or direct colleagues). In such a case the principles of a blind review may no longer hold as the chair has direct access to all reviews and the reviewers' identity. Similarly, it can also occur that a reviewer has submitted a paper and is thus able to discover the identity of the reviewers of his/her paper. In the latter case, reviewers can ask the chair to take appropriate measures.[11] In the former case, a reviewer may ask to be dispensed of reviewing the chair's paper or call upon the professionalism of the chair in case of a negative review.

[10] Some chairs or editors apply this trick as well in order to reach the minimum number of reviewers for a paper if there are not enough "original" reviews.

[11] For example, by adapting some settings of the reviewing system as not to disclose the identity of reviewers to the other reviewers.

3.2.2 What to Summarize?

Some review forms are unclear concerning the relevant form field "summary." It can be interpreted as a summary of the review or as a summary of the paper. The former case is meant to give an overall assessment of the paper, while the latter serves to show that the reviewer has understood the paper. In this case, do not simply copy/paste (sentences of) the abstract of the paper.

Example 20: (review summaries of a paper submitted to a high profile international conference)

Reviewer1: [overall: reject]

It is a theoretical paper in which many claims are done. It takes an X modelling approach for modelling A. I believe that B should be specified before modelling A, and not after, as the authors claims

Reviewer2: [overall: reject]

The paper raises a very interesting and relevant problem. Unfortunately, the proposed solution is not convincing and of low technical quality.

Reviewer3: [overall: weak reject]

The paper says too much about the problem and not enough about the solution.

The authors surely remember what they have written, so as a reviewer it is pointless repeating original sentences written by the authors to show that one has understood the paper. In fact, it only stresses the laziness of the reviewer and sheds doubts on the seriousness of the entire review. If a reviewer is not able to correctly formulate in his/her own words what a paper is about, an author may have justified reasons to doubt the appropriateness of the reviewer's comments and judgment. On the other hand, a reviewer might argue that the author(s) failed in presenting their ideas in an understandable way.

Anyhow, authors, even if reviewers have misunderstood the paper, can learn from the way reviewers summarize their paper as misinterpretations can point to inadequately formulated sentences, ill-structured sections, badly explained concepts, and so on. Therefore, reviewers should spend enough time and effort to produce their own summary of the submitted paper and include the summary of their review in the free text "detailed comments" field (unlike Example 20).

Example 21: (reviewer summary of a paper submitted to an international conference)

Reviewer3: [Recommendation: Accept]

This paper introduces an approach for a semantically enhanced A. While addressing a lack of current A technologies (referred to B in the paper) in facilitating the identification and association of A information, the authors outlined a prototypical work bench, which automatically identifies semantical

Rs between information items. This Q is explained by using a practical example taking the F, which is a R of a C in the US to file S, as the input for an A. In chapter 3 the methods for the identification and association of D items as well as the personal M are well explained using the concepts of a widely used L and a methodology for an X. Moreover the architecture is depicted with an example, how to achieve a semantically enhanced A via P, the use of Gs, and the mark up of semantical P belonging together.

This reviewer clearly made an effort to describe in a neutral way what he/she considers to be the most important points as well as the internal, logical structure of the submission. The author should feel confident that the contribution of the paper is well understood by the reviewer and will be more open toward critical remarks (unlike Example 20).

3.2.3 How to Organize Your Review?

Over time, a good reviewer develops his/her template to present the outcome of his/her evaluation. Make sure that your evaluation at least comprises the following items:

- Summary of the paper: one or two paragraphs summarizing the problem tackled, the approach proposed, the validation of the approach, and the main results reported.
- Positive issues or strong points: a list of the positive contributions of the paper.
- Negative issues or weak points: a list of the negative aspects of the paper or main drawbacks.
- Detailed comments: a detailed explanation of the strengths and weaknesses of the paper according to the evaluation criteria.
- Questions to the authors: a list of questions for the authors to reply to in order to define their proposed approach more precisely.
- Comments to the conference chairs or journal editors: a summary of the review and an overall recommendation.
- (Additional) comments raised after having read the authors' rebuttal responses and final decision (in case the paper has entered a rebuttal procedure).

Example 22: (complete reviewer comments on a paper submitted to a high profile international conference)

Reviewer3: [overall: weak reject]

Summary: *The paper proposes a solution for the semantic modeling of Y, their visualization, and sharing. The proposal is consistent with P and should increase the reusability of X founded B.*

Evaluation:

Paper strengths:

1. *The paper has clear objectives (i.e., propose a solution for committing, visualizing, and storing Bs) which are well embedded in current research, including D, O modeling, and X.*
2. *The paper is rich and has many contributions: representation and formalization of Bs, extending O to visualize Bs, extending the O for storing and sharing Bs.*
3. *As far as I can judge, the technical solutions offered are well-thought of and sound.*

Paper weaknesses:

1. *The writing of the paper is awkward at places and can be improved. Some sentences are hard to understand or difficult to interpret because of wrong choice of words or grammatical errors. The help of an English reviewer is advised.*
2. *The paper is not self-contained and I found it hard to understand without reading some of the previous works of the authors and their colleagues.*
3. *There is some inconsistent use of terminology. For instance in section 5 it is not clear whether L and R are synonyms or refer to different things.*
4. *The main weakness of the paper is the lack of an application. The authors do not demonstrate the benefits of O of Bs.*

Minor remarks:

Page 2, line 7: ?the those?
Page 2: incomplete sentence ?Since Bs are designed as declarative statements that constraint B in S [27].?
Page 3, second to last line: ?OderManager? instead of ?OrderManager?
Page 9, step 4: double use of ?Table 3?, second occurrence should be Table 4.

This reviewer has introduced a simple but clear structure in his/her review: first the summary, subsequently the strong and weak points, and finally some detailed comments on language and other minor issues. The review treats a variety of aspects. It starts with a summary to show that the reviewer grasps the essence of the submission. The reviewer has looked at the objectives of the submission as well as how these relate to the then state of the art. He/she has also examined the technical aspects of the proposed solution, including how the solution would be applied in practice. The structure, consistency, and overall presentation (including comprehensible language) of the submission are also addressed. Additionally, the review contains specific examples of "problems" that are well identified or located.

3.2.4 How to Meta-Review a Paper?

The meta-reviewing process involves senior reviewers to consolidate the evaluations given by the various reviewers of a paper and justify an acceptance or rejection decision (cf. Sect. 1.2.4). In many cases, evaluations by the reviewers may be divergent and even conflicting, so that a meta-reviewer has to moderate a discussion among the reviewers and guide them toward a consensus. Meta-reviewers should read all the papers assigned to them and provide a short evaluation comprising at least the following points:

1. A brief summary of the paper including strengths and weaknesses
2. An explanation of the most important criticisms of the reviewers
3. Comments on the response or rebuttal arguments of the authors
4. A decision of acceptance or rejection

Example 23: (complete meta-review of a submission to an international conference)

Meta-reviewer: [decision: rejection]

The paper proposes a twofold approach to solve the problem X. As highlighted by the reviewers, the solution is novel and has the potential to efficiently identify this problem. Furthermore, the authors addressed some of the comments given by the reviewers during the rebuttal phase. Nevertheless, the solution is poorly described and relevant problems are not treated at all (e.g., Y and Z). Additionally, the proposed approach is not compared with respect to state-of-the-art techniques (e.g., T1, T2, T3) and existing benchmarks (e.g., B1) are not used in the study. We encourage the authors to consider the reviewers' comments and resubmit the paper to another conference later this year. The recommendation is to reject the paper.

In Example 23, the meta-reviewer first summarizes the main topic treated in the paper and points out the most significant positive and negative comments of the reviewers. Then, the meta-reviewer assesses the quality of the rebuttal response. Finally, he/she presents the reasons that justify the rejection of the paper. Note that the meta-reviewer gives a positive message to the authors, namely, on the novelty of the approach.

Example 24: (complete meta-review of a submission to an international conference)

Meta-reviewer: [decision: acceptance]

The paper evaluates state-of-the-art approaches that provide efficient solutions to the X problem. As highlighted by the reviewers, the paper reports an extensive evaluation of a large variety of Y models, where different metrics are used to measure changes in the distributions of the indexed data and keys. The reported results reveal interesting properties of the studied models and can be used to advance the state of the art. We recommend the acceptance of this paper, provided that in the camera-ready version the authors follow the suggestion of the reviewers. Especially, the following issues must be addressed: minor typos in the definitions, motivation and interpretation of the studied metrics, and precise descriptions and discussions of the reported results

In both examples the meta-reviewer sends a clear message to the authors, facilitating their future work. Again, he/she gives a short summary and reiterates the strong and weak points of the paper as identified by the reviewers. In particular the meta-reviewer indicates which of the reviewers' suggestions the authors certainly have to take into account for the camera-ready version. Both meta-reviews follow the outline presented at the beginning of this section and consist of one or some sentences for each of the recommended points listed.

As an active researcher, you will eventually be invited to act as a meta-reviewer. Always keep in mind that authors deserve a clear and detailed evaluation of their work. Justify every criticism and try to give enough feedback to authors to help them to get their paper accepted in a next round of reviews.

3.3 To Conclude

This chapter included a collection of good and bad examples of paper review practice. For the sake of the argument, many bad examples have been included, which might give the wrong impression that many of the reviewers were not doing a good job. Luckily this is not the case and many reviewers hand in helpful and well-elaborated reviews. Nevertheless, the growing number of reviews to be done and the increasing importance given to high citation impact put a growing pressure on reviewers who become more and more tempted to economize on their reviewing time and effort. Some of the examples illustrate the results of this detrimental evolution.

Maintaining the peer review system requires many hours of intensive work by scientists around the world. Although the system is not perfect, as members of a scientific community we can help to overcome its drawbacks. Thus, as responsible reviewers we should ensure that authors receive helpful, fair, and unbiased evaluations resulting in a global advance of the state of the art. Only in this way, the process of peer reviewing, as one of the most important pillars of scientific quality control, will retain itself its high level of quality.

Chapter 4
Conclusion

This volume serves as an introduction to peer reviewing with the explicit goal of educating young or starting researchers in the subject. Peer reviewing is an important part of science and scientific activities as it functions as an internal quality control process. Despite this important role, hardly any structured training to peer reviewing is available, a lack this booklet aims to address.

Not only are the broad context, notion, and function of peer reviewing explained, but also the most general principles to adhere to have been presented as well as a lot of practical tips and tricks illustrated by concrete examples with explanations. Although more can be said about peer reviewing, in particular about its basic principles, and many more examples with accompanying explanations can be provided, a candidate reviewer should be well prepared for the job after having read this booklet.

In that sense, the material in this volume is well suited to be used in the context of innovative doctoral schools or other forms of training for Ph.D. students. And it is our hope that you, our reader, after having become an accomplished researcher and reviewer, pass on this message to your junior or starting colleagues by appropriately preparing them to become seasoned reviewers as well.

P. Spyns, M.-E. Vidal, *Scientific Peer Reviewing*,
DOI 10.1007/978-3-319-25084-7_4

Appendix: Selected Literature and Links

This appendix contains a collection of links and references to publications that are related to (the practical aspects of) peer reviewing. Many references come from the medical domain, but the general principles also apply for computer science. In fact, there seems to exist an overall shared and agreed-upon idea of what good peer reviewing should consist of.

All the links have been checked and accessed on 21 July 2015. It was not the intention to be exhaustive. A standard Internet search led to an initial list of texts which were kept if they showed some practical usefulness for starting reviewers. The bibliographies in this "seed list" were examined to collect additional articles that deal with peer reviewing in practice.

Some Papers on Reviewing in General

- _, (2015), Scholarly Communication and Peer review—the current landscape and future trends, The Wellcome Trust, http://www.wellcome.ac.uk/stellent/groups/corporatesite/@policy_communications/documents/web_document/wtp059003.pdf
- _, COST Action TD1306 New Frontiers of Peer Reviewing, http://www.pecrc.org/
- _, The Springer page on Publishing Ethics for Journals: A guide for Editors-in-Chief, Associate Editors, and Managing Editors, https://www.springer.com/gp/authors-editors/editors/publishing-ethics-for-journals/4176
- Dale Benos, Edlira Bashari, Jose Chaves et al. (2007) The Ups and Downs of Peer Review, Advances in Physiology Education 31: 145–152, http://

P. Spyns, M.-E. Vidal, *Scientific Peer Reviewing*,
DOI 10.1007/978-3-319-25084-7

depts.washington.edu/uwbri/PDF%20Files/Benosetal2007_AdvPhysiolEduc%7BUpsDownsofPR%7D.pdf
- LiquidPub Consortium (2009) Analysis of Reviews and Modelling of Reviewer Behaviour and Peer Review, in deliverable D.1.1 "State of the Art", pp. 29-58, http://disi.unitn.it/~birukou/publications/papers/2009LP_D1.1.pdf
- A. Mulligan, L. Hall, E. Raphael (2013) Peer Review in a Changing world: an International Study Measuring the Attitudes of Researchers. Journal of the American Society for Information Science and Technology 64: 132–161. Full report: Sense about Science Peer Review Survey 2009: http://www.senseaboutscience.org/pages/peer-review-survey-2009.html
- Azzurra Ragone, Katsiaryna Mirylenka, Fabio Casati, Maurizio Marchese (2013) On Peer Review in Computer Science: Analysis of its Effectiveness and Suggestions for Improvement. Scientometrics 97: 317–356, http://link.springer.com/article/10.1007%2Fs11192-013-1002-z
- Drummond Rennie (2003) Editorial Peer Review: its Development and Rationale. In: Fiona Godlee, Tom Jefferson (eds) Peer Review in Health Sciences, 2nd edn. BMJ Books, London, pp. 1–13 http://www.culik.com/1190fall2012/Paper_1_files/rennie.pdf
- Robert S. Turner (2009) Independent Peer Review Helps Assure Quality, Value, Objectivity. Journal of the National Grants Management Association 17: 43–48, http://projects.ecr.gov/moriversciencepanel/pdfs%5CBestPracticesPeerReview.pdf
- Richard Walker, Pascal Rocha da Silva (2015) Emerging Trends in Peer Review—a Survey. Frontiers in Neuroscience 9:169, http://journal.frontiersin.org/article/10.3389/fnins.2015.00169/full
- J. Wilson (ed) (2012) Peer review: The Nuts and Bolts. Standing up for Science 3, Sense about Science, London, http://www.senseaboutscience.org/pages/peer-review-the-nuts-and-bolts.html

Some Guidelines for Reviewers with Practical Suggestions

- Dale Benos, Kevin Kirk, John Hall (2003) How to Review a Paper. Advances in Physiology Education 27: 47–52, http://advan.physiology.org/content/27/2/47.long
- COPE (2013) Ethical Guidelines for Peer Reviewers, http://publicationethics.org/resources/guidelines

- Council of Science Editors (2012) Reviewer Roles and Responsibilities, http://www.councilscienceeditors.org/resource-library/editorial-policies/white-paper-on-publication-ethics/2-3-reviewer-roles-and-responsibilities/
- Frederic Hoppin Jr. (2002) How I Review an Original Scientific Article. American Journal of Respiratory and Critical Care Medicine, 166: 1019–1023, http://www.atsjournals.org/doi/pdf/10.1164/rccm.200204-324OE
- Brian Johnson (2015) Across the Desk: An Editor's Guide to Peer review Best Practice. http://exchanges.wiley.com/blog/2015/02/12/across-the-desk-an-editors-guide-to-peer-review-best-practice/
- Melina R. Kibbe (eds) A Primer for How to Peer Review a Manuscript for JSR, http://www.journalofsurgicalresearch.com/pb/assets/raw/Health%20Advance/journals/yjsre/JSR-Primer.pdf
- Chris MacQuarie (2012) Dear Buggy/Cher Bibitte. Bulletin of the Entomological Society of Canada, 44: 28–32, http://esc-sec.ca/bulletin/Bulletin_Mar_2012.pdf
- David Moher, Alejandro R. Jadad (2003) How to Peer Review a Manuscript. In: Tom Jefferson, Peer Review in Health Sciences, Wiley-Blackwell, pp. 183–190 http://www.bmj.com/sites/default/files/attachments/resources/2011/07/moher.pdf
- Nayera Moftah (2009) Reviewing Scientific Articles. The Journal of The Egyptian Women's Dermatologic Society 6: 3–8, http://www.jewds.eg.net/pdf/vol_6_1/2.pdf
- Kimberly Nicholas, Wendy Gordon (2011) A Quick Guide to Writing a Solid Peer Review. In: EOS, Transactions American Geophysical Union 92: 233–234, http://publications.agu.org/files/2012/12/PeerReview_Guide.pdf
- Ian Parberry (1989) A Guide for New Referees. Theoretical Computer Science, Information and Computing 112: 96–116, http://ac.els-cdn.com/S0890540184710534/1-s2.0-S0890540184710534-main.pdf?_tid=56c4b910-28b1-11e5-81de-00000aacb35d&acdnat=1436717963_4ebf39b14aa9a10a8bb767106e7d3dea
- Jigisha Patel (2015) A Beginner's Guide to Peer Review: Part Two, http://blogs.biomedcentral.com/bmcblog/2015/06/08/beginners-guide-peer-review-part-two/
- James Provenzale, Robert Stanley (2006) A Systematic Guide to Reviewing a Manuscript. Journal of Nuclear Medicine Technology 34: 92–99, http://tech.snmjournals.org/content/34/2/92.full
- Jennifer Raff (2013) How to Become Good at Peer Review: A guide for Young Scientists, http://violentmetaphors.com/2013/12/13/how-to-become-good-at-peer-review-a-guide-for-young-scientists/

- Jay Rojewski, Desirae Domenicon (2004) The Art and Politics of Peer Review. Journal of Career and Technical Education 20: 41–54, http://scholar.lib.vt.edu/ejournals/JCTE/v20n2/pdf/rojewski.pdf
- Alan Jay Smith (1990) The Task of a Referee, http://www-sop.inria.fr/members/Arnaud.Legout/Documents/external_reviewing.html
- Britta Teller (2014) Peer Review, https://brittzinator.wordpress.com/2014/10/17/peer-review/
- J. Winck, J. Fonseca, L. Azevedo, J. Wedzicha (2011) To Publish or Perish: How to Review a Manuscript. Revista Portuguesa de Pneumología 17: 96–103, http://www.elsevier.pt/en/revistas/revista-portuguesa-pneumologia-320/artigo/to-publish-or-perish-how-to-review-manuscript-90002033

An additional reviewing guidelines paper, but seen from the opposite view (what not to do) and written in a humoristic style.

- Graham Cormode (2008) How NOT to review a paper. The tools and techniques of the adversarial reviewer. SIGMOD Record 37: 100-104, http://www.sigmod.org/publications/sigmod-record/0812/p100.open.cormode.pdf

Albeit in a medical context, this paper reports on a tutorial on peer reviewing for undergraduate students.

- P. Rangachari (2010) Teaching undergraduates the process of peer review: learning by doing. Advances in Physiology Education 34: 137–144, http://advan.physiology.org/content/ajpadvan/34/3/137.full.pdf

Papers That Discuss Possible Future Trends in Peer Reviewing

- Richard P. Gabriel (2002) Writers' Workshops & The Work of Making Things. Addison Wesley Longman, New York, https://www.dreamsongs.com/Files/WritersWorkshop.pdf (prepublication version)
- Neil B. Harrison (1999) The Language of Shepherding. A Pattern Language for Shepherds and Sheep. In: Proceedings of the 6th Conference on Pattern Languages of Programs (PLoP 1999), http://www.europlop.net/sites/default/files/files/3_TheLanguageOfShepherding1.pdf
- Nikolaus Kriegeskorte, Diana Deca (eds) (2012) Beyond open access: visions for open evaluation of scientific papers by post-publication peer review, http://journal.frontiersin.org/researchtopic/beyond-open-access-

visions-for-open-evaluation-of-scientific-papers-by-post-publication-peer-review-137#articles
- Nikolaus Kriegeskorte, Alexander Walther, Diana Deca (2012) An emerging consensus for open evaluation: 18 visions for the future of scientific publishing. Frontiers in Computational Neuroscience, http://journal.frontiersin.org/article/10.3389/fncom.2012.00094/full

The following paper is interesting to see how the then future vision in 1999 has evolved in 2014 as in 1999 social media and related technologies still were at the very beginning of their later widespread adoption.

- Richard Smith (2003) The future of peer review. In: Peer Review in Health Sciences. Fiona Godlee, Tom Jefferson (eds), 2nd edn. BMJ Books, London, pp. 329–346, http://www.bmj.com/sites/default/files/attachments/resources/2011/07/smith.pdf

Examples of Bad Practice

An entertaining topic is that of (semi-) automatically generated bogus papers that have been submitted (and accepted!) after a so-called serious reviewing process:

- http://www.nature.com/news/publishers-withdraw-more-than-120-gibberish-papers-1.14763
- https://en.wikipedia.org/wiki/Sokal_affair
- http://marginalrevolution.com/marginalrevolution/2012/10/nonsense-paper-accepted-by-mathematics-journal.html
- http://www.theguardian.com/technology/shortcuts/2014/feb/26/how-computer-generated-fake-papers-flooding-academia
- http://smritiweb.com/navin/education-2/how-i-published-a-fake-paper-and-why-it-is-the-fault-of-our-education-system
- http://retractionwatch.com/2013/10/03/science-reporter-spoofs-hundreds-of-journals-with-a-fake-paper
- http://retractionwatch.com/2012/09/17/retraction-count-for-scientist-who-faked-emails-to-do-his-own-peer-review-grows-to-35/#more-9761

Printed in the United States
By Bookmasters